W9-ADO-135

Advance Praise for *Gender Medicine*

"There is, and always has been, a difference between men and women. Despite the fact that in our modern western society women adapt more and more to the attributes of men, the female body works differently. Marek Glezerman explains in an entertaining way the differences between male and female and its impact on the development, diagnostic and therapy of disease."

—Professor D. Wallwiener, President, The German Society of Obstetrics and Gynecology, Chairman, Dept. of Women's Health, University of Tübingen

"*Gender Medicine* underlines that gender medicine is not a corner or a niche for aficionados, but a mandatory revolution that has transcended all medical specialties. Marek Glezerman has written a comprehensive and absolutely unique book with a holistic view on human life from the angle of gender- and sex-specific medicine. It is not only most attractive for laypersons, but also very important for teaching students due to the discussion of a vast spectrum of gender differences in many areas of human health and disease, including the doctor-patient relationship."

—Giovanella Baggio, M.D., President, The Italian Society of Gender Medicine, Professor of Gender Medicine, Chief, Internal Medicine Unit, University of Padova, Italy

"Professor Glezerman's overview of gender medicine is a breakthrough in understanding the way we should practice modern medicine. A must-have for any health professional and an eye-opener for every inquisitive reader."

—Eyran Halpern, M.D., Chairman, the Israel Association of Hospital Directors, CEO, Rabin Medical Center, Israel

"Although not a traditional scientific composition, this fascinating, transformational book is a must-have for every physician, biomedical researcher, and medical student. *Gender Medicine* is at the core of the precision medicine concept that is changing the face of the medical field. Glezerman's comprehensive, multifaceted, and engaging approach to gender differences makes the case for the obvious, yet still neglected truth, that as much as a child is not a small adult, a woman is not just a version of a man."

—Professor Rivka Carmi, M.D., President of Ben-Gurion University of the Negev, former Dean at the Medical School of Ben-Gurion University

"*Gender Medicine* will lead to more involvement and dissemination among the general population and by stakeholders and decision makers."
—Karen Schenck-Gustafsson M.D., PhD, FESC, Professor of Cardiology, Founder and Chair, Centre for Gender Medicine, Karolinska Institute, Stockholm

"A pioneering, innovative book beautifully written by one of the foremost researchers on gender medicine. *Gender Medicine* is full of insights clarifying our past and projecting on our future. It posits the liberal principle of gender equality on its true basis: women are entitled to the medical treatment befitting their special traits as women. The vision portrayed in the book is compatible with the current shift towards personal based medicine whereby medical treatment should be predicated not on group statistics (the group being usually comprised of men only), but on the individual genetic traits of the patient, with gender being a major component. Undoubtedly this excellent book will be a torch of change towards gender medicine to which medical care should aspire."
—Professor Nili Cohen, President of Israel Academy of Sciences and Humanities, Professor of Law, former Rector of Tel Aviv University

"*Gender Medicine* comes at the right time. It is written in a clear and didactic manner, it provides the reader with a general overview, while offering ample reference material to dig deeper into specific areas of sex- and gender-specific medicine. I enjoyed reading the book and the sheer wealth of the information it contains."
—Professor E. Shalev, Dean, Rappaport Faculty of Medicine, Technion-Israel Institute of Technology, Chairman, Israel Ministry of Health—National Council of Obstetrics, Gynecology, Neonatology and Genetics

GENDER MEDICINE

The Groundbreaking New Science of Gender- and Sex-Related Diagnosis and Treatment

---

Marek Glezerman, M.D.

Foreword by Amos Oz

Overlook Duckworth
New York • London

This edition was first published in hardcover in the United States and United Kingdom in 2016 by Overlook Duckworth, Peter Mayer Publishers, Inc.

NEW YORK
141 Wooster Street
New York, NY 10012
www.overlookpress.com
For bulk and special sales, please contact sales@overlookny.com,
or write us at the address above.

LONDON
30 Calvin Street
London E1 6NW
info@duckworth-publishers.co.uk
www.ducknet.co.uk

Cataloging-in-Publication Data is available from the Library of Congress

Book design and type formatting by Bernard Schleifer
Manufactured in the United States of America
ISBN: 978-1-4683-1318-5 (US)
ISBN: 978-0-7156-5114-8 (UK)

FIRST EDITION
1 3 5 7 9 10 8 6 4 2

This book is dedicated to my closest family:
my dear friend and adored wife Zvia,
my beloved daughters Shira, Maya, and Tamar,
my beloved brother and dear friend Avi,
and my five grandchildren—great joys in my life.

CONTENTS

5: Gender Aspects of Heart Disease *95*

This chapter deals with gender aspects of the human heart and gender differences in diagnosis and treatment of heart disease.

6: The Digestive Tract—Gender Aspects *115*

In this chapter, the functional intricacies of the gastrointestinal system and the gender-related differences in the function of the various components of the gastrointestinal tract will be addressed.

7: Our Intestines—The Second Brain and the Life within the Gut *125*

Two major aspects of the gastrointestinal tract will be discussed at greater length: The first aspect relates to the microbioma, the huge population of bacteria that inhabits our intestines and with whom we live as a functional and symbiotic unit. The second is what has been termed "our second brain," namely the nervous system, which resides within the walls of our intestines and functions largely with little control by our brain.

8: Gender Aspects of the Reproductive System *141*

In this chapter, the differences in the reproductive systems will be discussed from the angle of divergent reproductive strategies, courtship, and the impact of differential reproductive processes. Finally, different types of disease of the reproductive organs in men and women will be discussed.

9: Gender Aspects of Infertility *153*

This chapter complements the previous chapter. Here we will discuss the differential therapeutic approaches related to men and women and also how men and women cope differently with infertility and infertility treatment.

10: Gender Aspects of Pain *167*

The basic concepts and the different types of pain will be discussed both in general terms and from the gender angle. Women perceive pain differently than men and react to certain pain medications differently than men.

Contents

FOREWORD
FIND THE DIFFERENCES
AMOS OZ

HANNAH, THE PROTAGONIST OF THE NOVEL *MY MICHAEL* (1968), tells us that her late father used to talk about men and women "as if the very existence of two distinct sexes was a disorder which multiplied agony in the world, a disorder whose results people must do everything in their power to mitigate."

How wrong he was, that old man! For is it not true that the very existence of two sexes that are so different is one of the most amazing gifts we have been granted, along with life itself, along with the joy of love and of parenting, along with the joy of creativity.

Yet generations upon generations of doctors, according to Marek Glezerman in his fascinating and innovative book, have disregarded this difference. Or to be more precise: generations upon generations of doctors saw women as nothing more than another type of men, perhaps a slightly defective form of men or a weaker form, but except for the matter of childbirth, basically identical to the "original."

Over time, and in many places even up to the present day, girls and women are treated using the same methods and the same medications and even the same dosages of medications as those used to treat men. The results are often damaging and sometimes even disastrous. This, Glezerman tells us, stems from the fact that over the years researchers and pharmaceutical companies tested new medications mainly on men, for among other things because the male body is not subject to the variations of the monthly menstrual period, thus making the experiments simpler and less expensive. All this may also be connected to a major trend in contemporary society which Glezerman does not touch upon in his book. I am referring to the fashionable tendency calling for modern women to dress like men, to adopt male mannerisms and habits, to become assimilated, at least to some extent, into a world shaped by men as a covert yet decisive condition for gaining status as an individual with equal rights in society and in one's place of employment.

The revolutionary concept of gender medicine, of which Glezerman is one of the proponents, pioneers and leaders in Israel and worldwide, stems in fact from the notion of human equality in its deepest sense and not its popular meaning: True equality between people does not mean that "we are all the same." Indeed, the opposite is the case. True equality means the equal right of each individual to be different, the equal right of each group to be different, the equal right of each gender to be different.

Glezerman's book is not written for researchers and scholars (though it is recommended reading for them as well). It is written for curious readers who love surprises and clarifying insights, readers who take pleasure in discovering new and provocative things about themselves, other people and the world around them. Marek Glezerman, like many other doctors before him, like Doctor Chekhov and Doctor Arthur Conan Doyle, knows how to tell an exciting story, a story full of surprises, full of ups and downs, a story whose heroes include strong and opinionated women who know how to communicate well and men who are strong to some extent and who learned many important things throughout evolution—though regarding how to communicate they would often do better learning from women. Perhaps first and foremost, before all the many insights this book provides us regarding men and women and regarding gender medicine, this book offers us profound discoveries regarding the links between body and soul, between disease and emotions, between pain and our ability to express pain, between the brain situated in our head and the opinionated and sophisticated second brain located in our digestive system. Isn't it true that we've always known that sometimes our thoughts come from our heads and sometimes from our guts? Glezerman's book confirms that this is indeed the case. Often our gut thinks without any involvement from our head and even without our being aware of it. (Today, right before our eyes we see how practically an entire nation, including its leaders, thinks directly from the gut and only from the gut . . .)

The book presents fascinating examples of the differences between the female brain and the male brain, differences that were shaped while we were still in our mother's womb. Similarly our sexual preferences also derive partially from the time we were in the womb.

Another chapter is firmly grounded in the new gender medicine, though its contents are indirectly yet clearly related to the works of Shakespeare and Virginia Woolf, Agnon and Dahlia Ravikovitch. This is the chapter on the gender aspects of pain. Who is more sensitive to pain, men or women? What are the types of pain that women are more sensitive to than men? During which periods of their lives? The chapter also includes discussions that rock the foundations of existing conventions and shatter prejudices (even those of doctors!) regarding the gender aspects of heart diseases and diseases of the reproductive system and the digestive system, and even the gender aspects of our life before birth – life as it develops in the womb.

The book also contains one sensational tidbit: the well-established and conclusive medical determination that men are in fact the weaker sex in the evolutionary kingdom. The male chromosome is undergoing a process of deterioration, and it is more than likely that men will totally disappear from the face of the earth within around two hundred thousand years. (We believe that this should force the United Nations into an urgent declaration that men are an endangered species. Though perhaps this is the case for the entire human race, if we base our judgment on what we see on television.)

All this science is described in *Gender Medicine*—in clear and fluent language, with humor, with a fascinating narrative style that causes the reader (at least this one) to longingly remember that good old-fashioned family physician, the family doctor from generations past, who had the time to talk to patients while treating them, to explain things in everyday language, to tell stories, to ponder out loud, to bring up fascinating memories and insights relying on fountains of wisdom and life experience. Yet not only does the doctor talking to us here in this book bring up memories but rather, and mainly, he tells us a new and revolutionary story about ourselves, a story that overturns many firmly established and ancient medical conventions and exposes us to new and provocative observations. And not only those about our own body. And not only those regarding medicine in the future. Much more than that. He spreads before us entirely new ways of looking at things from the social and cultural perspective and even from the political perspective.

The subtitle of the Hebrew version of this provocative book is *Toward the Obvious*. And rightly so. It has always been the case in science, and sometimes even outside of science, that what was unbelievable yesterday has today become a matter of curiosity, challenge and controversy before turning into what is obvious tomorrow.

I read this book almost like I would read a thriller. And like what often happens with thrillers, when I reached the end I said to myself: "I really knew this all along." The expression "gender medicine" is new. But the inner intuitive knowledge that there are a few things in this world that differentiate men from women is of course ancient, as old as humanity itself. Now, with the emergence of gender medicine and the publication of this fluent and articulate book, what we always felt and always assumed and always speculated has finally become obvious.

AUTHOR'S NOTE

THIS BOOK IS FOR ANYONE WHO IS CURIOUS TO LEARN MORE about the wonders of the human body. It is based on a lecture series I gave for laypeople in collaboration with Tel Aviv University, which was aired in 2013–2014. The series was later published in book form in Hebrew and has since been reprinted four times. For this English edition I have completely rewritten the book and have added five new chapters. In its present form, this book is still aimed at but is not limited to the lay reader, and will, I hope, benefit physicians, students, paramedical professionals, nurses, psychologists, physiotherapists, dieticians, and anyone else who deals with the diagnosis and treatment of men and women. In addition, I anticipate that every patient or potential patient will find interest in this book and come across at least one or two items which seem to have been written specifically for her or him. I have also included over 280 references and a thorough index to help readers dig more deeply into those areas which are of particular interest.

It is not necessary to read *Gender Medicine* from cover to cover, or even in order. Except for chapters 2 and 3, 8 and 9, and 12 and 13, which should be read successively, each chapter can be read as a stand-alone piece. The common theme underlying the book is based on the understanding that all bodily systems in men and women, although looking similar, may differ in the way they function and may respond differently to treatment, once they are diseased. While I've attempted to select the most important subjects for this book, given space constraints I have had to overlook others, including sleep, travel medicine, surgery, geriatrics, pediatrics, pulmonology, ophthalmology, oncology, dermatology, transplantation medicine, basic research, psychology, and sexology. These areas are of great interest to the field of Gender Medicine and will, I hope, be explored elsewhere.

A few words about words: For the sake of simplicity, I've used singular pronouns (his or hers) rather than referring to individuals in both the masculine and feminine form (she/he or his/hers). It should be obvious that whenever applicable, both sexes are meant. Likewise, throughout the book I will use the term Gender Medicine to refer to what should more accurately be termed Sex- and Gender-Based Medicine. Wherever I speak about the fetus, I do not distinguish between fetus and embryo and when I refer to the pregnant woman as "mother," I am aware that motherhood commences at delivery. Finally, whenever I speak of differences between the sexes, I do not mean that *all* men or *all* women differ in this or that characteristic. Overlaps exist to a higher or to a lesser degree and not all men or all women have all male or all female characteristics. I am also aware that many theories and hypotheses presented throughout this book may not be universally accepted and moreover that different theories or hypotheses may exist of which I am unaware. Such is medicine: we often come across different opinions and even different scientific data which seem to contradict each other but which may at the same time be correct. To detail such controversies would have been beyond the scope of this book. Gender Medicine deals with people who are treated by people, so there cannot be any absolute black-white, true-false fact.

GENDER MEDICINE

INTRODUCTION

WHAT IS GENDER MEDICINE?

DESPITE THE FACT THAT WE ARE INCREASINGLY AWARE OF specific gender/sex differences between women and men, the practice of medicine remains stubbornly stuck in the past. We regularly read about studies showing that women's symptoms—for example, when experiencing a heart attack—are quite different from those of men, and that ignorance of such differences has resulted in mortality and low-quality treatment. It's time to bring medicine into the twenty-first century with our new understanding of gender and sex differences, and I hope to help stimulate the conversation of how in this book. Some time ago I was asked about an opinion related to a young woman who suffered from repeated epileptic attacks which required repetitive changes in her medications. It turned out that both frequency and the intensity of her attacks were substantially higher during the second part of her menstrual cycle, which is among others characterized by secretion of the hormone progesterone. This hormone has the tendency of somewhat neutralizing antiepileptic substances. The appropriate treatment for this woman was therefore not to change medications, but to increase the dosage of her medications during this specific time window. Her neurologist accepted this suggestion and the problem was solved. This is an example of how understanding the physiological differences of men and women may successfully affect treatment approaches.

Human development on Earth began four to six million years ago. In cosmic terms, the history of mankind accounts for a mere fraction of a second. Even so, we have had millions of years to develop our genetic signature. Two million years passed before our ancestors developed the ability to use tools. It took another million years before we stood up on our hind legs to become *Homo erectus*—upright man. Neanderthals made their first appearance a million years later, and approximately 100,000 years ago, our species, *Homo sapiens* ("the wise man"), finally emerged.

By contrast to this long genetic history, all of human culture as we know it—the Bible, the Egyptian pyramids, philosophy, mathematics, medicine, agriculture, and the like—only began to develop in the last 5,000 years. It was twenty-five generations ago that we learned to use the compass effectively, which enabled us to conquer the seas and the oceans, and only twenty generations ago that we developed the mechanical clock, which enabled us to measure time and divide it up accurately. In (relatively) short order we invented the microscope, the printing press, the steam engine, and the automobile. And consider the revolutionary developments that can be attributed to human beings over the past eighty years, a mere two generations: the discoveries of penicillin and insulin that have saved the lives of millions of people, remarkable discoveries in the domains of agriculture, construction, mechanics, optics, transportation, aviation and the study of the universe—but also the development of weapons of mass destruction. And what about the most recent generation? Along this same exponential curve of inventions and innovations we have developed tools that have changed the world beyond recognition: the personal computer, the Internet, the cellular phone, social networks—all of these inventions have been developed in a single generation and together have created a whole new environment, vastly different from that of our less technologically advanced forebears.

What can we learn from this short overview of the history of humanity? First, human development for the most part was slow and—for a long period of time—linear. Our ancestors adapted as the planet's conditions, and particularly its climatic conditions, changed over millions of years. The Earth moved from a period of warming to an ice age and back again, and these changes led to the disappearance of certain species of prehistoric humans, while other species emerged and flourished. Humans developed the ability to stand on two feet, discovered the use of fire, domesticated animals, and invented agriculture. All of these events occurred over an extended time frame, ultimately improving human beings' ability to cope with their environment and driving major changes in our biology.

Yet while our biology has a long history, most of the dramatic changes in our way of life have taken place only during the last 100–150 generations, that is, over the last several thousand years. During this period human development began to grow exponentially, as Ray Kurzweil describes in his pioneering book, *The Singularity is Near*.[1] Kurzweil explains how human biology, which up until recently progressed very slowly, has been forced to cope with the dizzying pace of the technological developments occurring around us. This acceleration affects all areas of our lives, most notably our health. While developing ever-more advanced technologies to treat disease and augment our limited human abilities, we've collectively forgotten about the deep history of our bodies—and with it, one of the most basic factors underlying how our bodies work.

For most of our species' existence, we had sufficient time to adapt to our environment and develop attributes, skills, and abilities to help us survive in the face of natural threats and thrive. According to Darwin, those who managed to best adapt to their environment, to learn, to develop suitable skills and pass them on to their offspring were more likely to survive than those who did not. Over time these advantages were incorporated into the genetic structure of human beings and passed on to later generations.

What this means is that in spite of our modern trappings, our bodies bear imprints of our ancestors' past behavior. To understand our biology, we must first consider how our deep ancestors lived—in caves, as early as two million years ago—and how their behavior might influence our health today. Though we lack thorough records of prehistoric human cave life and thus must depend on hypotheses, assumptions, and scant evidence, there is strong reason to believe that the division of social roles—which was remarkably static over the course of human development—has resulted in significant physical and physiological differences between the two sexes. This difference, which is largely neglected by contemporary medicine, is what forms the basis for modern gender and sex-specific medicine.

We can assume that the need to divide the roles between men and women was related mainly to reproduction and to the social structure that emerged to support it. The instinct to have and raise children, to continue the species, was of prime importance. The women who gave birth, breast-fed, and raised the babies were naturally more confined to the cave or later to their permanent place of residence. They were especially vulnerable during and after their pregnancies and in need of protection. Their major role was therefore to get through pregnancy safely, raise offspring, look after the "nest," and contribute to the general economy by gathering food near their place of residence. The men were charged with protecting the group and obtaining food rich in protein—hunting game. Conforming to these roles necessitated differential development of men's and women's bodies and resulted in physical attributes that still prevail in modern humans, most notably, in size. In general men are larger than women and have around 20–30% more muscle mass. On average, men are 8–10% taller than women.

Today, the male advantage is size, height and strength could be considered superfluous, particularly since our social structure has changed significantly from prehistoric times. Indeed, one could argue the opposite: that some attributes of the male physical advantage are now more harmful than beneficial to societal functioning. The increased aggressiveness in men (stemming from their hormonal structure and physical strength) is one of the causes of violence in society in general and against women in particular. Most violent crimes are committed by men. Perhaps if masculinity did not find expression in greater physical strength and muscle mass, the problem of violence against women would be less prevalent. What was once essential for our existence seems sometimes not only unnecessary but detrimental to life today.

A less obvious holdover in male biology concerns pain tolerance. The hunter male developed unique abilities and attributes in order to increase the chances of survival for himself and his family. As hunters and warriors, men developed a higher tolerance to pain than women, a gender difference I'll discuss more in depth in chapter 11. Pain resistance was probably regarded as a sign of masculinity in ancient times, as it still is today. The biological basis for this higher pain tolerance among men is the male hormone testosterone.[2] The testosterone level gradually decreases after men reach their thirties, as does their tolerance to pain. Older men do not simply complain more about pain—they are actually more sensitive to it than they were when they were young.

Verbal ability also varied among men and women, which led to interesting implications for brain function. For men, speaking was a means of transmitting essential information. On a hunting expedition, not only was excessive talking unnecessary, it risked scaring off prey and attracting unwanted attention from predators and enemies. In contrast, women who set out together to gather food had more freedom to communicate. They transmitted information to one another about the plants they saw as they worked, asked questions, and also reported on possible dangers.

In the age of modern imaging technology, these differences take on new meaning. Functional MRI (fMRI) enables us to monitor brain activity under different conditions, such as when an individual is angry, sad, or laughing or during various cognitive activities to see which regions of the brain are activated. Researchers were surprised to discover that in performing identical activities, different regions are activated in the male brain than in the female brain. For example, during verbal activities a specific region in the left brain was activated for men, while for women different regions in both sides of the brain were activated.[3] This led them to conclude that men have a single speech center, while most women have two or more, perhaps offering a biological basis for the Talmudic saying: "Ten kabs of talk descended to the world: nine were taken by women." From the perspective of gender medicine, this difference is significant: when a woman has a stroke that affects her speech center, she responds better to treatment and thus her recovery is generally swifter than that of a man who has undergone a similar event.[4] The central importance of verbal ability in women may also be one of the reasons that girls start speaking earlier than boys do, and acquire a larger vocabulary. In general, verbal ability is more highly developed among women than among men, even in early childhood."[5,6,7]

The immune systems of men and women also differ, having assimilated ancient gender roles. For millions of years of human development, primary responsibility for raising the offspring, at least in early childhood, belonged to women, either singly or in groups. In addition, mothers continued to breast-feed their babies for as long as possible. For this reason, babies and young children were much closer physically to their mothers than to their fathers and this pattern remains consistent even today. (It is only in recent years and in certain population strata in developed countries that men have begun to take a more active role in raising their young children.)

In gender medicine, the fact that the task of raising children was historically relegated to women is of particular importance. The close physical contact necessitated by this role placed women at greater risk of contracting infectious diseases from their babies and children. Thus, as a form of protection they developed more robust immune systems than men did. Women enjoy the benefits of this inheritance to this day, as they are less vulnerable to infectious disease than men. Yet this is a mixed blessing. Women are also more susceptible to autoimmune diseases in which the immune system goes rogue and attacks the body it's meant to protect.[8] Today we know of more than 70 autoimmune diseases, most of which affect women far more than men. They include arthritis, which affects four times more women than men, autoimmune thyroid disease, which affects eight times more women than men, and lupus, which affects ten times more women than men.

The field of orthopedics offers other cases of gender differences leading to differences in health outcomes. Over the course of millions of years when women were responsible for raising their children, they developed improved fine motor skills in their hands, skills that are evident in women today as well. Most women have a significantly wider range of thumb joint movement than do men.[9] But again, while the advantage of these superior fine motor skills among today's women is not clear, women continue to pay a price for them. In women, arthritis often begins in the thumb joint.

Finally, women are more likely to bear knee injuries because of the size of their pelvis, which is necessary for giving birth. When the human species began walking on two feet around three million years ago, women were forced to develop a different stance than men in order to balance their weight. They do so by straightening their knees more than men when they stand up. This stance places more weight on the knees and as a result knee injuries are two to eight times more prevalent among women athletes than among their male counterparts.[10]

As the ones who raised the young, mothers also needed to be able to decode their babies' needs based on their facial expressions. Therefore, the ability to understand facial expressions in particular and interpret body language in general is more highly developed among women than among men[11] (see chapter 13). Female intuition is to a large extent based on these abilities which characterize women to this day. In the context of gender medicine, deciphering nonverbal language is important in doctor–patient communication (chapter 15). Such examples are few of the many I'll discuss in this book, which will show how the environment of prehistoric humans shaped today's bodies and continue to affect our health. Understanding this history is essential if we are to develop a medicine that can address and effectively treat the bodies of both men and women.

• • •

The central aim of gender medicine is to acknowledge the physiological and pathophysiological differences between men and women in the treatment of their bodies. Most of these differences, which developed over the course of millions of years of evolution, still exist today, largely without any real benefit. Yet the very existence of these differences requires that the medical profession awaken to the fact that men and women have different needs when it comes to their health.

We need to conduct more research on diseases that affect both sexes but which may manifest themselves differently in each. It is important to understand why certain diseases appear with varying frequencies among men and women and sometimes at different degrees of seriousness. We must study how medications affect the two sexes and to what extent these medications cause differing side effects among men and among women. We must learn how to apply the conclusions of animal studies to both human sexes.

The diagnostic process is also dependent upon the suitability of diagnostic tools to men and to women. In addition, the gender of the individual making the diagnosis affects the results. In light of all the above, gender medicine seeks to create new definitions of diseases and to spearhead the discovery of more accurate, specialized diagnostic and treatment methods for men and for women. Gender Medicine is important for all medical disciplines and many areas beyond. Actually, this new angle from which we look at human beings is of importance for every professional who takes care of women and men and for everyone who receives such care. I will therefore try to cover a rather large spectrum of topics from various areas starting off with a discussion about gender medicine and personalized medicine. Since gender differences originate in life before birth, I will devote two chapters to this topic with an additional chapter on the rather neglected issue of how stress on the pregnant woman may affect her unborn child and lead to future mental problems, often differently for males and females. I had to choose which bodily systems to discuss from the gender aspect. Each organ and system in our body, each medical discipline would have provided ample evidence for gender differences. I have picked the heart and the digestive systems but I could also have discussed others. No book on gender medicine can skip the reproductive system, which presents the most obvious gender-differentiated topic, so two chapters will deal with this. Four chapters are devoted to topics which transcend medical disciplines, including pain, temperature regulation, and gender aspects of the doctor–patient relationship. The latter topic also required a discussion about communication between people in general. Finally, and in order to emphasize that gender medicine is not merely upgraded gynecology, I have dedicated two chapters with provocative titles to the male.

So, before talking about gender differences we need to clarify the distinctions between gender medicine and personalized medicine, a topic which very often induces discussions. This is what the following chapter is about.

CHAPTER 1

SEX, GENDER, AND PERSONALIZED MEDICINE

AN ADMISSION I SHOULD MAKE UPFRONT: IN SPITE OF THIS book's title, the term "Gender Medicine" is not strictly accurate. I adopted the term for the sake of simplicity, and before we move on to a fuller exploration of medicine according to sex and gender, I'd like to first clarify the differences between these two interrelated but distinct concepts.

The idea of gender, borrowed from sociology, refers to a group of people in society, classified according to the group's unique characteristics, such as culture, social fabric, customs, behavior, values, and sex. Included under this classification of gender are the individual's role in society, his or her self-definitions and expectations from society, manner of dress, and numerous other parameters. The attributes of gender are fluid—they can change over time or in different locales. They depend not on biology, but rather upon the social setting in which an individual lives at a given point in time. In other words, gender is not something fixed that someone **has** but rather something that someone **does** or how he or she acts and functions in a given environment.[1]

Sex, by contrast, at least among human beings, is defined by biology and chromosomal structure. We differentiate between **genotypic** sex, or whether one has a Y chromosome or not, and **phenotypic** sex, which is the expression of this chromosomal structure and their genes in the individual's physical appearance and attributes. Yet other factors, such as epigenetics and hormonal processes, can also influence the way genes are expressed and which together generate one's phenotypic sex. (I'll expand on this subject later and in Chapter 3.)

In the animal world, sex is comparatively more fluid: a creature's sex is not necessarily determined by its chromosomes. The sex of turtle and crocodile offspring is determined by the surrounding temperature at the time the egg hatches. In turtles, eggs hatched at lower than 27 degrees Celsius leads to the birth of male turtles, while those hatched above this threshold will be female. Among crocodiles, the situation is reversed.[2] Certain fish dwelling in coral reefs can change sex from female to male or vice versa when prompted by an environmental catalyst, for instance if the dominant male of the group dies. This change includes the outer appearance, the sex organs, the sex glands, and the ability to produce sperm cells. The entire process takes only a few days.[3]

Although nature offers many such examples of sex being determined by environmental needs rather than by a fixed chromosomal structure, in most mammals, being male or female are fixed biological definitions. Changes in the chromosomal structure require far more than a death in the family— they require tiny evolutionary changes over the course of thousands and tens of thousands of years.

Biological sex may be hardwired, but as we are increasingly seeing with the growing visibility and equality of those who identify as LGBTQ (Lesbian, Gay, Bisexual, Transgender, and Questioning), as opposed to the hard-wired characteristics of the male and the female sex, the categorical division into masculinity and femininity is not biological and does not indicate core characteristics, but is rather based on society, education, social value systems, gender roles, and is given to rapid changes. For example, certain professions or ways of behavior can within a few years lose their masculine or feminine connotations. Just think of how quickly concepts of "feminine dress" are changing in the world of fashion or what masculine behavior means in various cultures and at different times within the same culture. The categories of sex and sexuality and the social subcategories of masculinity and femininity are not binary nor mutually exclusive. Biological and social categories exist along a continuum, and there is inevitable overlap between categories. In other words, the lines separating environmental and biological changes on the one hand and femininity and masculinity on the other are rather blurry. Gender Medicine—which attends to all these aspects under the umbrella of sex and gender—should more accurately be called **Sex- and Gender-Based Medicine.**

That out of the way, I'll explain how Gender Medicine addresses and incorporates these distinct factors to gain a fuller understanding of an individual patient. Diagram 1 describes the structure and areas of interest of Gender Medicine: At the base is the chromosomal: hardwired biology. On top of this is the next layer, representing the biological changes that men and women have undergone over millions of years of evolution. These biological adaptations allowed our ancestors to meet specific needs related to their socially defined gender roles— men as hunters and defenders, and women as gatherers and as those responsible for raising the children. These changes were passed on genetically from generation to generation, and this inheritance forms the basis for the major biological differences between the human sexes today.

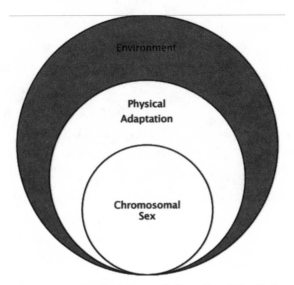

Diagram 1: Definition of Gender Medicine

Physical adaptation refers to the adaptation of the human body
to the needs of the roles human beings have performed
over the course of millions of years of evolution.
Environment refers to the adaptation of the human body
to the needs of the roles human beings play in the society
in which they currently live.

Finally, I've added an additional layer over the biological-chromosomal layer to represent the social-environmental layer, that is, the layer of gender according to its social definition. Each of the components of this structure is strongly connected to health and sickness, yet it's amazing that until now the field of medicine had focused virtually solely on only chromosomal sex in this context.

As I'll explain below, Gender Medicine takes into account the social-environmental aspect of gender as well as its biological aspect—both of which are essential to understanding how our bodies work (and break) and how to treat patients most effectively today.

ENVIRONMENTAL GENDER MEDICINE

Environmental Gender Medicine has to do with how gendered behavior and societal norms affect the bodies of women and men in very different ways. A few facts to illustrate what I mean:

- Women are prone to trachoma, an infectious disease that causes blindness. The difference in morbidity stems primarily from environmental conditions and gender roles. Trachoma is caused by a bacterium called chlamydia that is transmitted by flies. The flies are attracted to secretions around the eyes and mouths of children who live under poor hygienic conditions; there is a high prevalence of this disease in villages in the Sudan, for instance. The children contract the disease, and their mothers are infected due to their proximity to the children. Because caring for children is the sole responsibility of women in this region, the rate of contracting this disease among women is much higher than among men.

- Similarly, women are more likely to contract schistosomiasis, also known as bilharzia, a disease transmitted by a parasitic worm called schistosoma. This parasite, which ranges in size from 7–20 mm, lives in rivers and lakes, primarily in Africa and the Far East, and enters the body through the skin. After it enters the body it settles in the host's internal organs and can cause them to fail. Approximately 200 million people in the world suffer from this disease, most of them women. Why? In many regions struck with schistosomiasis, women are responsible for washing clothes and doing the cleaning. Consequently, they spend much time barefoot in rivers, putting themselves at greater risk of exposure to the parasite. Were local gender roles switched, with men spending more time in infested rivers, more men might be affected by schistosomiasis.

- Malaria, which is caused by a one-celled parasite transmitted by the Anopheles mosquito, is far more common among men. In the regions of the world where this mosquito is prevalent, cultural and religious norms dictate that women be covered from head to toe. They are thus more protected from mosquito bites than men in these regions, whose skin is generally more exposed.
- Carpal tunnel syndrome—a common repetitive stress injury caused by a pinched median nerve in the carpal tunnel in the wrist—is twice as prevalent among women as among men.[4,5] This is because jobs requiring repetitive use of the hand and wrists, such as waiting tables and especially typing, are mainly held by women. Just to give you an idea, if a typist types 400 strokes a minute, she will type around 190,000 strokes in an eight-hour workday. If 20 grams of force are required for each stroke, she will exert around four tons of force on her wrists over the course of a workday!

These examples show that we need to understand how the social fabric affects gender if we want to make effective changes to our environment in order to improve human health.

BIOLOGICAL GENDER MEDICINE

As important as the environmental aspects of gender medicine are, as a physician, I've chosen to focus primarily on the biological aspects of gender medicine in this book. The functional differences between men and women stem from differential evolution of the two sexes over the course of our deep history, a development that has been assimilated into our genetic and biological structure. This difference is expressed, for instance, in our cardiovascular system (Chapter 6), our digestive system (Chapter 7 and 8), the way we experience and respond to pain (Chapter 11), in our immune system, and how we process medications. Yet in spite of the knowledge already existing on this topic, most diseases and medications have been and still are researched and tested predominantly on men, as well as on male animals. For a select few diseases like osteoporosis, depression, and male breast cancer, the opposite is the case—we use data from women in order to make deductions about men. Drawing diagnostic conclusions about one sex based on clinical experiments on the other sex is like designing evening gowns for women using a mannequin of an average male body. It's time for a paradigm shift in medicine. It's time we turn to Gender Medicine.

GENDER MEDICINE VERSUS
PERSONALIZED MEDICINE

Before we begin our exploration of gender medicine, however, I'd like to address the topic that has recently dominated discussions about the future of health care: Personalized Medicine. For the first time in history, the raw materials of the human body have been deciphered, and the mapping of the human genome is enabling us to understand the recipe according to which the cells of the body grow and develop. This ability to analyze our bodies at a cellular and molecular level is likely to change the face of medicine beyond recognition. In many instances it is already possible to diagnose genetic conditions, to predict the development of genetically caused diseases down the line, to predict future fertility and the emergence of physical and mental illnesses, and to propose ways of preventing future illnesses based on personalized treatment or at least to suggest unique means for each individual to prepare for the emergence of certain illnesses.

In a number of fields, we are already witnessing the success of personalized medicine. For instance, in 2012, American researchers[6] reported on successfully deciphering the human fetal genome from the mother's blood in the first trimester of pregnancy. Using this technique it is possible to identify incompatibility between the mother's blood type and that of the fetus in order to take timely measures to prevent disease in the fetus and newborn without the need to treat all pregnant women in whom such an incompatibility is suspected based on their blood type only. Doctors and scientists have also had great success using personalized medicine in the field of oncology, or cancer treatment.

Faced with the seemingly unlimited potential unleashed by the deciphering of the human genome, as well as ever-decreasing costs of individual testing of the human genome, it's easy to let the imagination run wild. In theory, you could give a mere drop of saliva or blood at the laboratory and discover the existence of genes that, for example, might lead to the emergence of high blood pressure in the future. Specific medications could be adapted especially for your genetic structure, which are likely to avert the emergence of the disease while preventing most side effects. No longer would you need general checkups or expensive treatments based on trial and error. Doctors and patients would know everything and be able to do everything—and at a fraction of the cost—just by looking at our genome.

There might even be no need for doctor–patient encounters. All answers and recommendations could come directly from the laboratory. This option of "direct to consumer testing" already exists on the market for various purposes, for example matching couples according to their genome. A commercial Swiss company is already offering for a modest fee genetic compatibility matching in various fields, for example attraction between two members of a couple, areas of interest and even chances for successful reproduction. In an ideal world, every individual with a medical problem would be treated according to easily affordable, readily available and comprehensive information derived from her or his genome. All necessary forms of treatment, be they medical devices, medications, or interventional procedures would be tailor-made to this information and would be applied in a timely manner. It would no longer matter if the patient was a child or an octogenarian, whether white or black, whether male or female, nor would his or her medical history be of great relevance. So, you may ask, why bother developing gender medicine when science is marching in the direction of such an incredible technology? Why bother wasting time on developing other diagnostic and therapeutic models including gender medicine?

Every new technology has its limits, and so has personalized medicine. It's clear that this idealized vision of an easy, universal, one-time test that keeps us healthy for all time is far from reality.

First, most illness is not caused by genetic factors exclusively, but by the interaction of genes and the environment.

We must take into account epigenetic changes, or when the same genes are expressed in different ways.[7,8] What does this mean? The field of genetics is concerned with the structure of the genome and with changes in the genetic material, that is, in the DNA molecules. The field of epigenetics, in contrast, focuses on heredity that is not based on changes in the molecular spectrum of the DNA. Whereas structural changes in chromosomes and genes require thousands and ten thousands of years to occur, epigenetic processes can take place in a single generation. In mice, hundreds of genes have been found that manifest themselves differently in the tissues of males and those of females.[9] I'll talk more about epigenetics in chapters 3 and 4.

Second, the deciphering of the human genome does not necessarily mean that scientists know all of its implications. What is unknown is still far greater than what is known. Although we've learned how to read the genomic sequence, we are still working out the biological meaning of the "words" we read and the interconnections and relationships among them. While we may detect an anomaly in the gene sequences of our genome, more often than not, we are still in in the dark about its future clinical significance. Moreover, many diseases involve more than one gene. The possible combinations are countless. Looking at the genome still somewhat resembles looking into space. We are able to map and catalogue our galaxy, but what do we know about the features of individual stars?

Thirdly, the development of new technologies does not necessarily ensure the efficiency of its implementation or its accessibility to all. Antibiotics were discovered more than eighty years ago. Yet, millions of people still die each year from easily treatable infectious diseases, in most cases due to accessibility barriers to medications. Forty years ago, monoclonal antibodies for cancer began to be developed, that is, antibodies developed from a single cell of a specific type of cancer.[10] Scientists predicted that these antibodies could be used as a basis for unique drugs that could be sent to attack and destroy the specific target cells, from which the antibodies derived. Many liked to think in terms of a "letter bomb" designed to be delivered to specific targets. Great hopes were placed in this development, and victory in the war against cancer seemed to be around the corner. And yet, even though this method proved to be very effective for various diagnostic and therapeutic approaches (in the United States each year around thirty monoclonal antibodies are sold at a value of twenty billion dollars), the technology has still not managed to overcome cancer and is not very likely to do so in the foreseeable future for many reasons. In the meantime we continue to diagnose and treat cancer with increasing success using other means as well.

Fourthly, there are serious ethical considerations related to Personalized Medicine[11,12] and controversy as to how to handle the large body of sensitive information collected from the genetic code of individuals. In the era of information explosion, privacy has become a major concern not only in politics and business, but also in medicine and health care. The potential harm of an individual's information about his or her health, illnesses, and chances of becoming ill in the future becoming available to government authorities, employers, romantic partners, and insurance companies, is immeasurable.

Additionally, we must cope with the question of how to appropriately use such information, once obtained. Checking for a mismatch of blood type between mother and fetus is a directed genetic test with clear therapeutic implications. What about diagnoses of diseases with no known cure? Or discovering that a fetus is liable to develop a certain illness in forty years, one which may or may not be treatable in four decades' time? How should the future parents act in such a case? How should we respond to uncertain information regarding unclear risks? In prenatal diagnosis: To what extent should we aspire for perfection when it comes to human beings? When and how should a couple determine to what extent imperfection is not deserving of life? The intention here is not to question a woman's right to abort her fetus according to certain norms, laws, and limitations. Yet it's undeniable that in the context of genetic testing, complex ethical issues arise. The results of fetal genome survey testing are generally not accompanied by the option of diagnostic procedures and are liable to confront parents with the difficult decision of whether to terminate the pregnancy or to wait to see whether the prediction will come true. This is a heavy emotional load to bear.

Finally, given the constant update of genomic information, how will medical professionals ensure such that applicable new discoveries will be called to patients' attention? Let us say I have undergone testing of my genome and the information is now in my medical file at my doctor's office. Information is constantly developing, and in another two years new information is revealed about the human genome indicating an increased risk for a certain future disease, a piece of information that affects me directly. What mechanism will be used for me being continuously updated related to my genome? Where will this information be stored? What new discoveries should I be informed about, and who should be informing me? And what if I am not interested at all in being updated with information that is not unequivocal? And what about data security? What about the costs? Not to mention the emotional burden on a healthy individual who worries on a daily basis about new emerging information related to his or her medical future. Will every patient want to follow these information curves like the performance of stocks in a portfolio? If science has developed the technological ability to provide information about the fate of an individual's health, that individual should be allowed to receive it if he or she so desires and to use it in certain instances based on professional advice. But how this should be done is far from clear at this time.

In contrast, Gender Medicine does not raise such ethical problems. Indeed the opposite is the case. Ethical questions emerge if medicine does not take the aspect of gender into consideration. But, should gender medicine be considered as an intermediate and temporary step towards personalized medicine? The answer to both questions is No. Personalized medicine is doubtless an important stage in the development of medicine, but it cannot replace everything that currently exists. It constitutes a new component in a comprehensive medicine based on facts, and indeed, medicine should integrate genetic information about the patient in addition to the medical and family history, the results of physical examinations, and the results of auxiliary tests.

Yet medical care should also take into consideration the patient's value system, lifestyle and habits, environmental factors, and of course age and sex. This is where gender medicine comes in. The goal is to draw a more complete picture that reflects the unique nature of individual people and their bodies. Based on all this, doctors can put together a unique and comprehensive diagnostic and treatment plan for each individual.

Personalized medicine cannot be compared to gender medicine. They operate on different ethical and technological planes, and their objectives are different. Personalized medicine is one of the greatest technical revolutions in modern medicine, but it cannot take the place of medical professionalism and medical acumen. It will always be easier to just look at your patient and consider obvious characteristics, which are important for your decision-making process, like age, size, race and yes, gender. Most medical services will continue to be based on diagnostic procedures and therapeutic approaches which are known today and on those which are going to be developed in the future. Gender medicine does not compete with personalized medicine, nor is it an intermediate stage on the road to the sophisticated technology of personalized medicine. It is not a technology at all, but rather a broader understanding of the medical importance of the distinctions between the sexes. It seems obvious, yet such a view has been ignored by the medical establishment for far too long. The rest of this book aims to address this deficiency.

I'll begin at the beginning: how does gender affect how we develop in the womb?

CHAPTER 2

LIFE IN THE WOMB—PART 1

I STILL VIVIDLY RECALL THAT NIGHT SHIFT ALMOST FORTY YEARS AGO. How proud I felt. A young on-call physician at the beginning of my obstetric residency, I had diagnosed that a first-time mother—to her great surprise!—was about to give birth to twins, and I had managed to deliver them successfully. It was only when I was stitching up the new mother's episiotomy that I noticed the presence of a third fetus. So, this night I was given the chance to deliver healthy triplets.

How little we knew then about life in the womb. The developing baby was hidden from our eyes and thus from our understanding as well. According to prevailing medical wisdom, the developing fetus was the ultimate embodiment of self-centeredness. It was thought to prosper in a protected environment, unaware of and isolated from the outside world. We assumed that the fetus took everything it needed for its development without regard for its mother's needs. For the mother, her growing baby remained in the realm of mystery, and even its sex was not known to her until its birth. Moreover, she was ignorant of how her lifestyle, daily routine, and diet could affect the baby—for better or for worse. Until the 90s, pregnant women smoked, drank alcohol, and took medications without being aware of how this might affect their fetuses.

The fact that women and society in general were unmindful about the exposure of the growing fetus to the external world is surprising: from time immemorial, people have believed that a pregnant woman's experiences influenced the development of her unborn child. Almost 2,500 years ago the Greek physician Hippocrates wrote that a woman's emotional state would affect her baby's appearance. The ancient Greeks believed that if a mother looked at beautiful objects during pregnancy she would give birth to a beautiful child, whereas if she looked at ugly things this would result in an ugly baby. They believed that if a pregnant woman was angry at someone, her child would resemble the object of her anger, and that a pregnant woman who looked at the moon would give birth to a child with psychological problems. (Indeed, the origin of the word "lunacy" is "luna," the Latin word for moon.) Others believed that cleft lip was caused by a mother who drank from a cracked glass and intelligence was due to the mother's reading habits. It was thought that if a mother drank wine during her pregnancy, her child would be born with what is called today a "port-wine stain"—a birthmark that looks like maroon wine was spilled or splashed on the skin. Such superstitions testify to the intuitive recognition that what a pregnant woman experiences could be passed on to her fetus and would affect it both cognitively and physically. This intuition, however, did not translate into any systematic, scientific adjustment of the pregnant woman's habits to best support her growing fetus.

Over the last several decades, dramatic changes have taken place in this domain. Thanks to the development of ultrasound technology, we are able to observe the fetus while it is still in the uterus and to diagnose a variety of pathologies and treat them (or at least to be prepared to treat them immediately after birth). The fetus has become a patient, and the medical discipline of obstetrics has matured into maternal-fetal medicine. More than forty years ago, Ian Donald, father of the modern ultrasound, predicted that "the first 40 weeks of existence may well prove to be far more important medically than the next 40 years."[1] Only now are we beginning to understand the enormous impact of these prophetic words, to appreciate the process of fetal programming, and—most important for our purposes—become aware of the developmental differences between male and female fetuses. The impersonal term *fetal programming*, adopted from the computer world, would appear to contradict the notion that the fetus already carries all its unique genetic fabric and needs no further "programming." Yet the term reflects the idea that it is not the genetic makeup of an individual alone which determines the expression of his genes, but a complex process.

Fetal programming is a relatively new term in medicine, based on the understanding that the developing fetus's environment has a tremendous influence on its later health, not only in childhood but well into adult life. Research has suggested that the intrauterine environment can affect set points for factors such as body weight and temperature as well as the functioning of physical and psychological systems. For the rest of its life the individual's body will try to achieve equilibrium according to these set points, which can have ramifications that reach far into subsequent generations.

While new in the medical lexicon, the idea of fetal programming is not novel. Aldous Huxley's futuristic 1932 novel *Brave New World* contains a hair-raising description of the practice of preparing embryos for the work—and life station—to which they will be assigned.[2] In the novel, the Director of Hatcheries explains how embryos destined to work as laborers in tropical climates are vaccinated for equatorial diseases in the hatchery. Embryos designated for work in chemical plants are exposed to lead and other poisonous substances; those ordained to become aerospace engineers are raised in vibrating test tubes; and those designed to work under particularly difficult conditions without asking questions are programmed to low mental capacity, stupefied by oxygen deprivation and exposure to alcohol. Putting aside the dubious morality (and legality) of such practices, it's clear Huxley was decades ahead of his time in his intuition regarding test-tube fertilization methods, a process known today as in vitro fertilization (IVF), and of the effects of the environment on the developing fetus.

GENETICS AND EPIGENETICS, GENOTYPE AND PHENOTYPE

Before we go any further, however, it's essential to first understand the basics of how genes are passed down and how they work.

A genotype is the genetic blueprint that one inherits from one's father and mother. A human being has approximately 20,000 to 30,000 genes in his genome, and genes that have not undergone mutation are passed on "as is" to subsequent generations. Changes in genetic structures and in gene activity require thousands of years. It took a long time for the human genome to continuously change over the course of millions of years in order to adapt, for example, to climatic changes on Earth.

One of the opposites of evolution is revolution—a fundamental change that takes place rapidly. When change is required within one generation, as I will discuss in the next chapter, it cannot occur over the course of a long evolutionary process on the genotypic level but rather must happen on the level of how genes (generally for physical and psychological attributes) are expressed. To examine these swift changes to the original "operating instructions" of the genes, we turn to epigenetics.

Epigenetics, from the Greek prefix *epi-*, meaning "nearby" or "above," is one of the mechanisms that can lead to rapid change. These molecular and biochemical processes do not change the sequence our DNA, but do change the way the genes are expressed. They can act like a switch to change, activate, or deactivate the operation of certain genes, or they can interfere with various corrective mechanisms of the genetic material. Epigenetic changes occur through one or more of three processes: histone modifications, micro RNA changes, and DNA methylation. The observation in mice of increased levels of DNA methylation of female placentas, may explain partly why females enjoy a better protection from exogenous insults than male fetuses.[3]

There are many examples of sex-specific epigenetic changes that inform our understanding of gender medicine. One such change relates to the hormone Leptin—also called the "satiety hormone," as it inhibits hunger. If the Leptin receptors in the placenta are epigenetically silenced by DNA methylation, male babies are born more lethargic and hypotonic, whereas female newborns are not affected.[4] In a sense, epigenetics is the tool that permits the environment to effectively modify the genetic blueprint of an organism without tinkering with the blueprint itself. Thus, epigenetic changes in the placenta which change gene expression will affect fetal development.

The phenotype is the result of epigenetic "individual interpretations" of existing genetic "instructions," including physical, behavioral, and psychological attributes. If the genotype is the root than the phenotype is the plant as it grows dependent on the environment. From the fetal programming perspective, the environment of the mother's womb has an enormous potential impact on epigenetic processes and is therefore crucial in determining the phenotype. To complicate matters, epigenetic changes can be temporary or irreversible and may be passed on to further generations.[5] Consider for example monozygotic twins which are created when an embryo splits into two. Both embryos contain the same genetic load. These twins are known as identical twins, and frequently they look identical, but not always. In some cases identical twins hardly resemble each other. This difference in appearance is usually the result of an epigenetic process which one or both twins underwent in the womb. In other words it is ultimately not the genotype alone that exerts control over our appearance and health, but rather the way the genotype is expressed. This expression can be influenced by factors both inside and outside the mother's uterus. Today we can control these environments, at least to some extent.

BOY OR GIRL? THE ROLE OF HORMONES AND GENES IN DETERMINING SEX

From 11 to 11.25 days after fertilization, one of the most critical windows of opportunity in the life of the mouse fetus opens for a period of six hours.[6] These six hours determine whether the developing fetus will have a male or a female phenotype. The fetus's chromosomal gender was already determined at the time of fertilization, but phenotypically, all options are open. If nothing happens during this window, the default is for the fetus to develop as a female regardless of its chromosomal structure, including the development of ovaries. This is similar in humans, although the precise analogous timing for humans has not as yet been determined. The mechanism and the genes involved in this process are considered to be similar in all mammals.

If the male Y chromosome is present and activated, the sex gland of the fetus—primitive at first—will develop into a testicle and begin to secrete large amounts of the male hormone testosterone. The fetus will then begin to develop into a male. The gene responsible for determining the male sex is called SRY because of its location on the Y chromosome (Sex Determining Region on the Y Chromosome).[7] In mice this gene can be detected for a period of two days at most. In contrast, it can be detected in testicular cells throughout the lifetime of the human male. Increased production of testosterone in the testicles of the fetus will have a decisive influence on the development of sex differences; unless the SRY gene activates another gene known as SOX9, the default for the sex gland is to develop into an ovary. Still, the ovary is not completely passive in this regard. It seems that the ovary also "wants" to develop spontaneously into a testicle but is blocked from doing so by a gene called FOXL2, which suppresses this impulse.[8]

According to the Bible, God created Adam first and later created Eve. Yet, based on genetic analysis and the development of the ovaries and the testicles, the order was most likely the opposite. Without the Y chromosome and the activation of its SRY gene, the human fetus will develop into a female. In this sense, the male might be defined as a modification of the female.

Of course, it would be an oversimplification to say that the development of a male fetus is dependent solely upon the hormonal environment. The importance of genetics in determining cell function in both sexes was memorably demonstrated in the study of a rare gynandromorphic finch.[9] Gynandromorphic means that the phenotype includes characteristics of both sexes (from the Greek: *gyne* = female; *andro* = male). What is indeed special about this bird is that its right brain and the right half of its body are genetically male, while its left brain and the left side of its body are genetically female (Figure 1). The feathers on the right side of the bird's body have a male appearance, while those on its left appear female. Moreover, the ovary on the left side of the bird functions properly, while there is a testicle on the right side, albeit a poorly functional one. The reproductive behavior of this finch resembles that of a normal male, though it cannot reproduce. Another notable feature relates to the bird's song circuits. Neurological song circuits are crucially important attributes in birds, as the type and manner of a bird's song have a major impact on its ability to find a reproductive partner. Accordingly, they are different not only for disparate species but different for males and females of the same species. The song circuits on both sides of this finch's brain are more masculine than those of normal female finches.

What happened to this finch during its development? It must have been exposed as a fetus to a male hormonal environment. Otherwise it would not have developed any male characteristics at all. But the fact that this finch retained not only female characteristics but also female genetics to such an extent indicates that other genetic forces must have been at work. In the case of this finch it appears as if the cells of the body, including those of the brain, had "remembered" their genetic structure despite the influence of the male hormone. Thus, even though there is no doubt that the environment in which a fetus develops has a determining influence on the behavioral and cognitive attributes of the entire phenotype, genetics plays an important role in determining phenotypic sex, too.

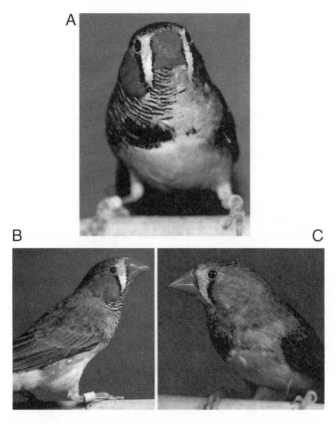

Figure 1: Gynandromorphic finch

Reprinted with permission. Agate et al, 2003.
Copyright (2003) National Academy of Sciences, USA.

Another illustration of this developmental complexity can be seen in fetal growth rate. The growth rate of the fetus is dependent on its hormonal environment but is also controlled genetically. If fetal growth were solely dependent on intrauterine testosterone levels, male fetuses would quickly surpass female fetuses in growth velocity once the testes began to function (around the seventh week of pregnancy). Yet we have shown in about 900 consecutive IVF pregnancies using sonographic measurements that male fetuses are larger in early pregnancies *before* the expected onset of intrauterine testosterone secretion.[10] This points to the existence of a significant genetic factor in growth before the onset of hormonal changes.

HORMONES IN THE WOMB

Once the sex gland has begun to develop due to the activation or non-activation of the SRY gene, male and female fetuses embark on paths down different developmental tracks. The male sex gland, the testicle, begins to secrete testosterone during the ninth week of human gestation. (Gestational age in humans is counted from the first day of the mother's last menstrual period, so the ninth gestational week corresponds to the seventh week after conception.) Testosterone levels reach their peak around the 14th to the 18th week of pregnancy, achieving levels close to those of an adult male. Testosterone has various effects on the developing organism, for example on the development of the sex organs and on fetal growth rate.

Testosterone's effects on the development of a fetus's sex organs have been known for many years. In 1959, researchers in the USA injected testosterone into pregnant guinea pigs, which caused the fetuses to develop male sex organs regardless of chromosomal makeup.[11] Smaller amounts of testosterone led to male reproductive behavior among female guinea pigs after birth.

Testosterone has a significant influence on the development of the mammalian brain, and specifically on the differences between male and female brains. The results of this development will have a determining effect on "male" behavior after birth and throughout life and can affect, for example, how a child behaves during play. Researchers from Sweden have demonstrated that girls with Congenital Adrenal Hyperplasia

(CAH), which is associated with high prenatal levels of testosterone preferred masculine toys, such as trucks. They also found a direct correlation between testosterone levels and preference of boys' toys.[12]

The influence of androgens (male hormones) *in utero* on the development of neurological and behavioral characteristics has been unequivocally proven in animal studies. Testosterone may even be involved in the establishment of sexual orientation and sexual identity. High testosterone levels in female embryos (or newborns) will often lead to behavior patterns typical of males. On the other hand, neonatal castration in male rats leads to typical female behavior. The critical period of time in which testosterone affects an individual's future development differs from species to species. In apes and in humans, the critical window of opportunity opens up before birth, while in rats it occurs immediately after birth. The effects, however, are similar.

Hormone levels in the uterus also seem to affect mental and behavioral characteristics of an individual after delivery. Researchers in England obtained and stored amniotic fluid from pregnant women and observed the behavioral patterns of their babies for up to 48 months after delivery. They measured the male hormone testosterone in the amniotic fluid samples and correlated the levels of fetal testosterone to the babies' subsequent development.[13] At the age of 12 months, researchers found that girls made much more eye contact than did boys and that eye contact—a sign of social communication—decreased as the level of fetal testosterone increased. The researchers continued to investigate the development of this effect between the ages of 18 and 24 months and discovered that girls had a much larger vocabulary than boys, with a significant inverse correlation between fetal testosterone and vocabulary. That is, as fetal testosterone increased, vocabulary level decreased.

Fetal programming relates to virtually all aspects of future life and reaches beyond health and disease. One example is fetal programming for financial risk-taking in future life. This should actually not come as a surprise. Intrauterine exposure to testosterone is causative to the development of the "male" brain, and this is associated with risky behavior of males throughout their life. Therefore, intrauterine testosterone levels could also be associated with an individual's willingness to take specific types of risk in his professional life. In a study of 550 MBA students at the University of Chicago,[14] researchers compared willingness to take financial risks to assumed levels of testosterone to which the students had been exposed in their mothers' wombs. For example, students were asked about the dollar amount they would be willing to pay in order to avoid participating in a 50/50 lottery that pays either $0 or $200. Since the participants' intrauterine hormone levels were not available to the researchers, they used two indirect measures.

The first was the ratio between the length of the second (index) finger and the fourth (ring) finger. Many studies have shown that the length of the ring finger is affected by intrauterine testosterone levels. As intrauterine testosterone levels increase, the ring finger grows longer and the ratio between it and the index finger decreases. In general, among men this ratio is smaller than among women. It is not entirely clear why the ring finger is affected by testosterone, and this theory has been the subject of some criticism. Nonetheless, this indicator may be used to deduce intrauterine testosterone levels.

The second indirect measure was a test used to examine an individual's ability to interpret someone else's emotions by looking at him. A participant is shown 36 pictures of eye expressions and asked to choose among suggestions of different emotional expressions, such as anger, passion, fear, and others. The test was originally developed to diagnose the autism spectrum, but is used today to test empathic ability. It seems that fetuses that were exposed to higher levels of intrauterine testosterone score lower on this test.[15]

To support the indirect estimates of intrauterine testosterone levels the researchers measured actual testosterone levels in the participants as well. The results showed that in general women were more risk-averse than men. It seemed that as levels of testosterone increased, so too did willingness to take financial risks among women and among men.

MOTHER-CHILD MESSAGING

One of the most important aspects of intrauterine programming is how the fetus responds to information it receives about the world outside the womb. Based on "messages" transmitted by the mother about the external world, the fetus adapts its physiology. Such epigenetic mechanisms compensate for the slowness of genetic changes, preparing the baby for the specific environment into which it is about to be born. If the messaging from mother to fetus conveys accurate information, the newborn will be properly equipped for the challenges that will in fact arise in his or her life.

In contrast, if there is no correspondence between an anticipated challenge and the adaptive tools developed by the fetus, it might find itself at a disadvantage. For example, maternal stress during pregnancy, resulting in the transfer of stress hormones to the fetus, may subsequently lead to increased vigilance and higher awareness of potential threats—i.e., anxiety—

in offspring. This can be appropriate if the outside environment is indeed full of danger. But if the newborn is born into a calm, protected home, the baby's suspicions regarding its surroundings together with rapid changes in attention may find expression in attention disorders and even in paranoia (see chapter 5).

Another example of inappropriate tools for the environment is in the case of obesity. If a mother-to-be transmits to her baby that the outside world is a difficult place with limited access to food, the fetus is programmed to hoard calories at any opportunity. If this baby is in fact born into an environment where food is scarce, it will have the tools to accumulate calories and food and thus will have a survival advantage. If, however, food is readily available, the baby will be at high risk for developing obesity (see next chapter). To corroborate this hypothesis, researchers from the USA limited the food intake of pregnant sheep.[16] When the newborn lambs were given free access to food, they overate and developed obesity. Their mothers' limited food intake may have transmitted to the fetuses a signal of limited access to food after birth. But when this limitation of intrauterine food supply continued after birth as well, the development of the lambs was similar to that of lambs whose mothers' food intake was not restricted during pregnancy. Hence, when a fetus's expectations about the outside world match the future environment, it leads to a survival advantage; when a mismatch occurs, it leads to a higher risk of disease.

This accurate or erroneous information transmitted by the mother is apparently based on her weight and weight gain during pregnancy. From the perspective of the fetus, it is irrelevant whether the mother's lack of appropriate weight gain during pregnancy derives from hunger or a self-imposed diet. Reduced calorie intake by his mother indicates reduced availability of calories and this sets the stage for epigenetic adjustment. Therefore it is important that weight gain during pregnancy remains within the recommended guidelines and does not deviate in either direction.

Another interesting example of maternal messaging involves a mouse-like rodent. Voles born in the fall have thicker fur than those born in the spring. While the thicker fur at delivery confers no immediate survival advantage, it's protective for the growing pup as winter draws near. Thus being born with thicker fur before the cold season serves as a kind of pre-birth investment for the future. Researchers found that the day length to which voles are exposed during gestation affects the mother's secretion of the hormone melatonin. This is transmitted as a message to the fetuses and affects how much fur they grow.[17]

In a more applicable example, food preferences in humans can be affected by fetal programming. As certain flavors in the food of a pregnant mother pass into the amniotic fluid, the fetus is exposed to some of its mother's taste preferences. Since food preferences are one of the characteristics of a cultural environment, the mother conveys to the fetus information about her cultural environment, that is the environment into which the baby is going to be born. In fact, studies have shown that babies preferred flavors to which their mothers had been exposed during pregnancy.[18] Moreover, studies indicate that fetal learning does take place, including gaining familiarity with the mother's speech characteristics and musical preferences. In these sophisticated ways mothers-to-be transmit messages to their babies, preparing them for the culture into which they are about to be born.

FETAL PROGRAMMING
AS SURVIVAL STRATEGY

Fetal programming has implications that reach far beyond preparing an individual fetus for adult life. From a broader perspective, fetal programming plays a role in ensuring the survival of the entire human species. In populations that have experienced extreme conditions such as hunger, natural disasters, or war, the gender ratio of newborns seems to skew in favor of girls. This was the case after the 1995 earthquake in Kobe, Japan, and the disasters that followed. Scientists hypothesized that this decrease in the gender ratio—the decrease in the number of boys born relative to the number of girls—occurred as a result of an increase in the frequency of spontaneous abortions involving male fetuses among women under major stress. Already forty years ago, Trivers and Willard,[19] the researchers who investigated this finding, suspected that from the perspective of evolution, there was some logic in spontaneously aborting weak male fetuses whose ability to reproduce is inferior to that of weak girls, thus enabling the woman to begin a new pregnancy that will lead to the birth of a girl or a more robust boy. This observation has also been confirmed in subsequent studies.[20]

This so-called Trivers-Willard effect (TWE) has been applied in a study of the sex ratio among offspring of members of the Forbes Billionaires list.[21] As predicted by the TWE, in the highest economic brackets there is a significant sex ratio bias in favor of sons, indicating that given a sound economic safety net mothers' bodies take the risk of raising "weaker" offspring. Data from a mathematical analysis conducted in the United States on a database of Swedish births in the period from 1751 through 1912 clearly support this hypothesis.[22] The researchers noted that the life expectancy of males born in years in which the sexual ratio favored females was longer. They took this finding to mean that once a selection mechanism had been initiated which preferred strong female offspring, male fetuses survived only if they were exceptionally robust. In this context, fetal programming may present an extreme Darwinian mechanism of "selective feticide" in a harsh environment: only the fittest survive. If this assumption is correct, it means that under extreme conditions intrauterine target values for survival ability are generated and male fetuses below these targets become at greater risk of spontaneous abortion.

It may be that the increased rate of spontaneously aborted males derives from the mothers' increased susceptibility under stressful conditions. Or perhaps the maternal system intervenes in the intrauterine environment, leading to deliberate spontaneous abortion of fetuses in accordance with their sex and their robustness. According to this theory, males born into a difficult environment should be more robust because the birth of weaker males is blocked by the mechanism of spontaneous abortion.

INTRAUTERINE ENVIRONMENT AND IQ

What about cognitive development? If maternal communication can lead to physiological adjustment of the fetus, does that hold for the brain as well?

The long-standing—and at times, fierce—nature-versus-nurture controversy has often focused on the heritability of IQ, a fertile ground for infamously racist theories and ideologies. However, in order to arrive at meaningful and scientifically sound results, we must take into account not only genetics, but the different environments in which a human being and his/her

IQ develops. Home, social, and educational frameworks are often-cited examples of such environments, but, I would argue, so is the uterus.

In 1997, psychiatrist Bernard Devlin and his colleagues[23] published a meta-analysis of 212 IQ studies which included over 50,000 twins. They reported that IQ is positively related to birth weight, suggesting that maternal nutrition may affect the IQ of her child. Moreover, they summarized literature indicating that IQ may also be increased by certain dietary supplements used by pregnant women, and lowered by the consumption of alcohol, drugs, and tobacco during pregnancy. Thus, the intrauterine environment as well as the postnatal environment affects an individual's IQ. The authors estimated that the total effect of genes on IQ is less than 50%. Moreover, the fetal environment alone accounts for 20% of the difference between twins and 5% between siblings born from the same mother at different times.

In another study[24] that included close to 60,000 children, psychology professor Eric Turkheimer showed that the relative contribution of environment and genetics to future IQ levels is closely related to the socioeconomic status of the family. The lower the status, the higher the impact of environment, and vice versa. According to this body of data, it seems that in the low socioeconomic bracket virtually all the variance of IQ is related to the pre- and postnatal environment while in the highest socioeconomic environment most of the IQ can be attributed to genetics.

DIFFERENCES BETWEEN THE MALE
AND THE FEMALE BRAIN

The differential development of the male and the female brain has been a fascinating topic of research for the past hundred years. These differences are not concerned (at least in today's science) with intelligence but with function, as I will explain.

The consensus today is that the secretion of testosterone in the fetus plays an important role in cell differentiation, as shown in both animal and human studies.[25] And sex hormones in the uterus to a large extent dictate the developmental differentiation of brain structures. In the human brain, sex-based differences have been reported in various regions, such as the hypothalamus, cortex, and amygdala. The development of fMRI (functional Magnetic Resonance Imaging) technology has made it possible to associate cognitive functions with specific brain regions. As a result, the brain, the most complex organ in the human body, has become more accessible for study.

Until recently, the assumption was that once the brain became fully developed, it was no longer capable of generating new cells. Today there is evidence that in different regions of human (and ape) brains, new nerve cells are generated throughout life. This phenomenon, known as brain plasticity, includes both anatomical and functional changes in response to external stimulation and sheds light on the brain's sensitivity to input.[26] In 2010, Dan Agin, a renowned American geneticist, summarized this by saying that ". . . from the perspective of brain neuroscience, the environmental world of the developing embryo and fetus is as important as any information by the genes . . ."[27] Agin concluded that the human brain is not a static and analytic system, but rather a dynamic system that adjusts in accordance with the input it receives.

Many, though not all, activity centers in the brain are located in only one brain hemisphere. The differences between the sexes in brain sidedness of functional regions has aroused much scientific interest, particularly findings regarding language. Many studies have provided evidence that a number of tasks related to language tend to be focused on the left side of the brain among men, while they are distributed between the two sides of the brain among women.[28] The two hemispheres develop at different rates, with the development of the right hemisphere preceding that of the left, and this may be the reason that language skills develop earlier in girls. Moreover, a certain gene (FOXP2) which is crucial for the development of speech and language has been found in the brains of 4-year old boys to be significantly lower that in girls of the same age.[29] High levels of fetal testosterone in the male result in a smaller corpus callosum—the main channel for transmitting information between the two hemispheres—and thus in lessened connectivity between the right and the left brains. This may be why it is easier for women to use both sides of the brain and why they are better at multitasking than men.

Researchers in the United States who examined nearly one thousand young people found that the male brain has more connections **within** each hemisphere, while the female brain has more connections **between** the two hemispheres. That is, adjacent regions in the brain are more connected among men, while distant regions are more connected among women. The researchers use this to explain why the male brain facilitates the association between perception and coordinated actions, the "control centers" of which are located in the same hemisphere, while the female brain facilitates communication between analytic and intuitive processes, the "control centers" of which are in both hemispheres.[30]

I'm using the terms "female brain" and "male brain" here for simplification—they do not imply a dichotomy similar to female and male sexual organs. There is extensive overlap between morphologic characteristics of the brains of women and men. Yet from a functional perspective, the brains are different. This notion has recently been disputed by Daphna Joel at Tel-Aviv University and her colleagues.[31] The authors examined data sets of MRIs of thousands of men and women and arrived at the conclusion that there is no such thing as a male and a female brain. This conclusion disregards the functional differences in the brains of men and women on grounds of quantitative measurements of various brain regions.[32]

It seems futile to try and distinguish between male and female features of the human brain by looking at structure instead of looking at function. That would be like looking at an elaborate road map and drawing conclusions related to traffic patterns. The brains of women and men are indeed different. Not better, not worse, not stronger, not weaker—just different. For one thing, the basic building blocks, the brain cells, differ chromosomally: The sex chromosomes of women comprise two XX chromosomes and those of men have a XY chromosomal pattern. The male brain is exposed to a completely different hormonal environment in the womb than the female brain. Overwhelming scientific data as to the crucial effect of testosterone on the developing male brain cannot seriously be challenged by morphometric imaging studies. Ovaries and testes function differently (see chapter 9), because of differences in hormone secretion by distinct brain regions (hypothalamus and pituitary). In the female brain this secretion is periodic, and that is the reason for the existence of the female menstrual cycle. In the male brain the secretion is almost constant and that is one of the reasons for the continuous process of sperm production. These brain regions look the same and the hormones secreted are exactly the same, but the pattern of secretion is different, leading to fundamentally different functions of our gonads. The effects of testosterone to which the male fetus is exposed (but not the female) are innumerous, including behavioral characteristics after birth, preferences for toys, verbality, communication, and others.

. . .

The impact of fetal programming on sickness and health throughout our lives cannot be exaggerated. Fortunately, major efforts are being invested in scientific research—notably in the United States—to collect as much data as possible on this topic.* The more we can learn about fetal programming the more we can help identify potential risks, learn how to avoid them, and cope with risk factors once they have been identified.

*One example is Project VIVA, a groundbreaking longitudinal research study funded primarily by the National Institutes of Health (NIH) and the U.S. Centers for Disease Control and Prevention (CDC). Launched in 1999, Project VIVA has been monitoring more than 2,500 children starting from life in utero. To date, over one hundred original research papers have been published as part of this project, in addition to book chapters and editorials. In 2003 another, even more comprehensive research project was launched in the United States, called the National Children's Study (NCS). The stated objective of this project was to recruit one hundred thousand women before and during pregnancy and to examine environmental and genetic influences on the growth, development and health of their children before birth and up to the age of 21. The study has been mandated by law and funded by the United States Congress and government agencies, and to date more than $600 Million US out of $1.3 billion US which was allocated originally to the study has been invested. Unfortunately, the project had to be put on hold for feasibility problems in December 2014 after enrollment of 6,000 children in 40 locations across the USA. Over 125 peer reviews studies were published based on the data accumulated.

In China, calculating a person's age is a complicated affair, based on a "nominal" count as well as an "actual" count. The general principle is that the age of a baby at birth is not zero but one year old, meaning that the first year of life is counted as one. Likewise, the concept of fetal programming counts the antenatal time period as part of the life of an individual, in fact, as the most important part. This new science reflects an understanding that our lives do not begin at delivery—at zero—but that our lives before birth are of crucial importance to our future health.

CHAPTER 3

LIFE IN THE WOMB—PART 2

L IFE BEFORE BIRTH IS NOT EXACTLY A PARADISE. THE ENVIRON-
ment in which the pregnant mother lives directly influences the developing fetus. Indeed, a more appropriate
way to describe the fetus in the womb is to liken it to an inmate
in a prison: It lives in captivity, it can neither fight nor flight,
and it is vulnerable to influences from the outside world as well
as to changes in the mother's physical and emotional systems.
In effect, there is no other period of life in which humans are
so susceptible to the environment without being able to influence or escape from it. In this chapter I'll explore fetal programming and its dark side—the effects of toxins such as
alcohol and tobacco on the developing fetus, and the fetal origins of disease—as well as how gender affects such effects.

THE PLACENTA

In humans as in other mammals communication between the
mother and fetus is mediated through a sophisticated time-
limited organ that develops precisely for this purpose and withers
shortly after the fetus is born. This is the placenta.

Following fertilization of the ovum by a sperm cell, the
embryo forms and cell division begins. After a series of cell divisions, a cystic structure is formed, the so-called blastocyst. This
structure subsequently implants on the inner layer of the uterus,

or the endometrium. By this point the blastocyst is already a differentiated structure. It includes an inner cell mass, which will develop into the fetus and an outer cell lining, which will develop into the placenta. Except for the uterine lining, which is of maternal origin and will become a layer called decidua, the entire placenta is genetically identical to the fetus, i.e., it can be female (46 xx) or male (46 xy). This is why a biopsy of the placenta in the first trimester of pregnancy can supply us with genetic information related to the fetus.

The human placenta is primarily responsible for transferring nutrients and gases for fetal development, removing waste, and mediating hormones and immune responses. Whatever the mother ingests or experiences may through the placental blood flow affect her developing fetus, including medications she takes, diseases, nutrition habits, and stress. Changes in the placenta as a result of such variables may be macroscopic, such as a reduction in placental size in smokers, or microscopic, such as changes of the lining of blood vessels in women with high blood pressure. Moreover, subtle changes in placental gene expression may impair signal transmissions to the fetus and specific placental enzymes may act as "sensors" for nutrients which respond to maternal cues. All of these processes may be related to epigenetic mechanisms.[1]

Any unhealthy habits or diseases of the pregnant mother will undoubtedly have an impact on the fetus, and in many cases this impact is quite grave. Toxic substances, including alcohol, nicotine, and pollutants, can cause harm to the newborn with implications for the rest of its life. The effect of external toxins on the fetus is dependent upon the type and amount of toxic exposure as well as on the stage of its development. Exposure to toxins during the first trimester of pregnancy often result in malformations or the death of the fetus, while exposure in the second and third trimesters can cause delays in prenatal development, growth restrictions, and failures of various bodily systems.

"EXTERNAL" INFLUENCES ON THE INTRAUTERINE ENVIRONMENT

Our living environment is replete with toxins. This is particularly true for women of low socioeconomic status. A 2002 study of the use of residential pesticides, mostly for cockroach control, among pregnant women in low-income neighborhoods in New York City[2] revealed that more than 85% of them had used pesticides in their homes during their pregnancies. All of the women monitored had detectable levels of three insecticides in their blood. These insecticides are neural toxins known to harm the developing fetus and are likely to cause permanent brain damage leading to cognitive deficiencies throughout the child's life.

Harm caused by prenatal exposure to toxins can also be passed on to later generations by epigenetic processes, as has been observed in rats exposed to chemical toxins before birth. Their offspring became ill with numerous diseases, including cancer, and these effects could be observed over the course of three generations.[3] There is also a social aspect to this. Since living in poverty increases the risks for detrimental epigenetic changes in the unborn and people are most likely to bear children with others within their socioeconomic group, poverty-driven pathologies may become hereditary.

Alcohol is a prime example of how the uterine environment can have a serious and deleterious impact upon the developing fetus. Up to one percent of all newborns in the United States each year, or approximately forty thousand children, are born with fetal alcohol spectrum disorders, a cluster of disorders that includes physical, behavioral, emotional factors, and learning disabilities, among others.[4] The severe end of these disorders is known as Fetal Alcohol Syndrome (FAS). Each year in the United States, between 4,000 and 12,000 children are born with this devastating condition. The syndrome is characterized by a severe delay in growth and disruptions within the central nervous system, as well as by structural defects and neurological, functional, and cognitive disturbances accompanied by peculiar and anomalous facial features. The prevalence of FAS is estimated between one and three in every thousand live births in the United States. This number is similar to the number of babies born each year with Down Syndrome, though in the case of FAS the tragedy could have been totally avoided. It is believed that alcohol consumption during pregnancy is the primary cause of mental retardation in the United States, responsible for more than all malformations due to chromosomal defects together.[5]

Unfortunately, the maximum permitted amount of alcohol intake after fertilization and during the course of pregnancy is not known. In view of this, the United States Department of Health and Human Services has unequivocally declared that "no amount of alcohol consumption can be considered safe during pregnancy."[6] There is no longer any doubt that a pregnant woman who consumes alcohol is endangering her fetus. Especially dangerous is the consumption of alcohol during the first weeks of pregnancy, when she may not yet know that she is pregnant, but when the fetus is much more sensitive to these toxins. Therefore, it seems advisable that women stop consuming alcohol in any amount not only once they know they are pregnant but as soon as they make the decision to become pregnant.

From the gender medicine perspective, male fetuses are more vulnerable to alcohol consumption. One of the reasons for this discrepancy is genetic. Mothers pass a variation of a certain gene (deiodinase-III) on to boys but not to girls. A study on rats found that this gene increases sensitivity to alcohol by disrupting the hormonal balance of the thyroid gland in the fetal brain.[7] The case of FAS in male fetuses illustrates the damaging impact of a combination of genetic hereditary processes (transference of a defective gene from mother to fetus) and epigenetic processes (changes in genetic expression due to the influence of alcohol).

Another well-known adverse impact on the fetus is the use of tobacco. If a mother smokes while pregnant she hinders the development of her fetus, resulting in a decrease in placental transfer of various important substances, low birth weight, and an entire spectrum of injuries to the fetus and newborn.[8] In utero exposure to nicotine may also be related to the onset of multiple diseases throughout the child's life, and these may differ depending on the sex of the child. In male rats, nicotine exposure in the womb causes high blood pressure in adulthood. Exposure to tobacco smoke during pregnancy increases the prevalence of cognitive problems and hearing impediments among offspring when they reach maturity, apparently due to a decrease in the thickness of the cerebral cortex— a region in the brain responsible for higher-level cognitive functions.[9] These injuries are more prevalent among mature female rats.

Other negative influences of smoking are uniquely related to gender. A study conducted on smokers and non-smokers with and without exposure to the mother's smoking during pregnancy found that exposure to nicotine as a fetus is related to reduced biochemical components in the brain that control attentiveness. Women more often exhibited a reduction in visual attention, while among men the reduction was primarily in hearing attention.[10]

"INTERNAL" INFLUENCES ON
THE INTRAUTERINE ENVIRONMENT

No less important than external influences are the "internal" effects, such as antibodies and hormones produced by the mother and conveyed to the fetus, and other substances that the fetus itself produces. A prime example is testosterone, which the testicles of the male fetus produce in large quantities.

Increased levels of testosterone and its derivatives have been observed in fetuses with an adrenal tumor or due to a genetic mutation of women that causes their children to be born with Congenital Adrenal Hyperplasia (CAH), a defect of their adrenal gland. In order to pass on this mutation to offspring, both parents, who have normal functioning adrenal glands, need to be carriers. One variation of CAH results from mutations that prevent the fetal adrenal gland from converting fetal testosterone into estrogen. As a result, the testosterone level in female fetuses with CAH rises dramatically toward the end of the first trimester of pregnancy. These female babies are then born with indeterminate or somewhat male-looking genitalia. After birth, the condition can cause early sexual development, obesity, excessive hair growth, as well as later absence of menstrual periods, and fertility problems. Even though medical treatment can usually solve this hormonal imbalance after delivery, this defect can still remain expressed in various types of behavior typical of boys, for example in preferences for "masculine" toys like trucks and weapons. Even after receiving appropriate treatment, affected women have a greater tendency to be sexually attracted to women than do women who were not influenced by overexposure to testosterone during their fetal life.[11] Furthermore, girls exposed to increased levels of testosterone in utero exhibit during adolescence a decrease in verbal ability, improved spatial orientation abilities, and heightened aggression—characteristics usually attributed to phenotypic men.

Women who suffer from polycystic ovary syndrome (PCOS), the leading cause of ovulation failure, have increased testosterone levels. Once pregnant, these women still secrete elevated levels of testosterone, which may affect their female fetuses. In a recent study it has been shown that offspring of rats in whom a PCOS phenotype was induced displayed more anxiety than the control group.[12]

On the other side of the spectrum is Androgen Insensitivity Syndrome (AIS), a condition in which someone who is genetically male looks and acts like a female. This syndrome is characterized by a normal male genotype and normal secretion of the male hormone testosterone, but inactive testosterone receptors. In other words, the AIS individual's bodily systems fail to "recognize" testosterone, which leads to ineffective intrauterine activity of this hormone. As a result, a female phenotype develops in fetuses that are genetic males. Very rarely is this syndrome identified at birth because the babies born look like females and develop as females throughout their lives, including their behavior patterns, their gender identity, and their sexual preference. Yet those with AIS have neither a uterus nor ovaries, and do not have menstrual periods. Absence of menstrual bleeding is usually the first reason that these girls seek medical help. The diagnosis is simple and relies on a genetic test and ultrasound which confirms the absence of ovaries and uterus. The external genitals usually look female with a functional, relatively short but blind-ending vagina. The testes are usually located within the abdominal cavity, where they originate. Since there is an increased risk for malignant development of these testes, most physicians advise surgical removal after puberty. These patients typically continue their life as normal but infertile women.

These two syndromes, CAH in female fetuses and AIS in male fetuses, clearly point to the decisive role of androgen levels and their intrauterine receptor activity in determining subsequent physical, behavioral, and cognitive development of men and women.

Another hormonal influence on the male fetus may be overexposure to testosterone prior to birth, which has been associated with the development of dyslexia. It has also been claimed that autism may be associated with high intrauterine levels of testosterone and that this impairment may be an extreme form of the "male brain."[13] Indeed, autism is four times more prevalent in men than in women and Asperger's Syndrome is nine times more prevalent in boys than in girls.

On the other hand, low levels of fetal testosterone has been detected in male fetuses when the mother is under severe and chronic emotional pressure, for instance during wartime, or when she is exposed to natural or personal disasters. Such conditions are liable to have a major impact on the development of the fetus (see chapter 5).

It's evident that the environment of the mother's womb has a strong impact on the fetus's subsequent cognitive ability and it has been suggested that[14] the impact of intrauterine life on subsequent intelligence equals or exceeds the impact of education. Just as important is the environment's effect on the fetus's health later in life, as I'll describe next.

FETAL ORIGINS OF ADULT DISEASE

It has become increasingly clear that many diseases in adults may originate in fetal life and early childhood. Obesity is one prominent example. Obesity has been declared one of the main risk factors in a variety of diseases and affects over a third of the adult population in the United States. In Israel, around 30% of adults are overweight and 15% are obese. In addition, around 8% of Israeli children between the ages of 6 and 14 are obese.[15.]

Extreme birth weight, whether too low or too high, is liable to lead to obesity later in life and to the emergence of metabolic syndrome, a cluster of factors that includes elevated blood pressure, abdominal obesity, elevated fasting plasma glucose and high triglycerides (a type of fat in your blood). The syndrome is characterized by the appearance of at least three of these components in a moderate form. Each component constitutes on its own a risk factor for certain diseases, and in combination, they raise the risk of major disease significantly.

Chronic diseases like osteoporosis, mood disorders, and psychiatric syndromes as well as polycystic ovary syndrome (PCOS) have been related to fetal programming. Other adult diseases such as cardiovascular disease, a variety of metabolic and endocrine pathologies, diabetes, and various others may also be due to inadequate determination of set points during fetal programming—our next topic.

MALNUTRITION OF THE MOTHER
AND THE FETUS

In 2004 the World Health Organization reported that more than 20 million children are annually born with low birth weight (less than 2,500 grams), 95% of them in third world countries. The report's authors agree that "Birth weight is a strong indicator not only of a birth mother's health and nutritional status but also a newborn's chances for survival, growth, long-term health and psychosocial development."[16] This is a strong statement, linking low birth weight*—whether due to premature delivery, growth retardation, or otherwise—to the future health of the newborn.

The concept of a relationship between fetal undernutrition and subsequent development of adult disease was introduced over two decades ago by epidemiologist David J. Barker.[17] The Barker theory, also known as the fetal origin hypothesis, states that low birth weight reflects intrauterine malnutrition which "programs" the fetus for later cardiovascular disease, diabetes, and hypertension. Specifically, Barker hypothesized that the fetus is programmed in response to messages from the mother that permanently determine target values for its metabolism throughout life. This hypothesis is one of the basic assumptions in research on fetal programming. Today, there is broad consensus that maternal nutrition has an impact on fetal health and that malnutrition of the mother, and particularly if her diet is low in protein, is associated with many illnesses that her fetus is liable to develop as an adult.

---

*In obstetrics we differentiate between intrauterine growth retardation (IUGR) or SGA (small for gestational age) if the fetal weight is below what should be expected at a given gestational age. If the birth weight is low because of premature delivery, we define this situation as Appropriate for Gestational Age (AGA). In both situations the end result is Low Birth Weight (LBW). Thus, the weight of a prematurely delivered baby may be either AGA or SGA. Notably, both AGA and SGA in preterm delivery may be major risk factors for later eating disturbances. Of course, it is not the size of the newborn per se which is the basis of pathophysiological processes in the future of this child. The reduced body weight may represent pathological processes occurring within a detrimental intrauterine environment, including placental insufficiency and epigenetic changes.

During World War II, the Allied Forces attempted to invade Holland, which was under Nazi occupation. The attempt failed, and as punishment the Germans imposed an almost total blockade on northern Holland from November 1944 through the end of May 1945. This blockade together with the harsh winter led to what later became known as the Dutch famine or the Hongerwinter ("Hunger winter"). Food rations dropped to 400 calories per day, or 25% of the minimum daily requirement. Many years later, a wide-ranging research study was launched at the Academic Medical Centre in Amsterdam to examine the effects of the Hongerwinter on men and women born in Amsterdam between November 1943 and February 1947, that is, prior to, during, and after the blockade. Thanks to carefully kept records, the researchers discovered that the birth weight of people who had been exposed to malnutrition as fetuses was lower by an average of 200–300 grams than that of the birth weight of fetuses of women who were not pregnant during the Hongerwinter.

Barker observed that these Hongerwinter babies were over the course of their lifetimes at high risk for a variety of adverse conditions, among them increased blood fats, obesity, diabetes, cardiovascular diseases, hypertension (high blood pressure), pulmonary diseases, and renal diseases. Hypertension is arguably one of the most deadly silent killers. It affects over 10% of the world population and is the major cause for cardiovascular disease and associated mortality. The majority of people with hypertension live in developing countries where medical care is restricted. The increase in cases of hypertension can be attributed to lifestyle change and reduced physical activity but, as in the case of the babies born after the Hongerwinter, also to insufficient maternal diet.

Moreover, it seems that intrauterine malnutrition affects

male and female fetuses differently.[18] Unlike male rats, female rats exposed to intrauterine malnutrition developed obesity and pre-diabetes as adults. On the other hand, human boys are generally more sensitive than girls to intrauterine malnutrition, among other reasons because of their more rapid growth rate and greater dependence upon sufficient nutrition. A woman carrying a boy in her uterus requires a caloric intake 10% greater than that required by a woman carrying a girl.[19] Low birth weight is also related to defective functioning of the endothelium during childhood. The endothelium is the tissue that lines the interior surface of blood vessels and lymphatic vessels.[20] This finding represents an important risk factor for the development of cardiovascular conditions. The discovery that endothelial damage in adulthood may originate during fetal life suggests that atherosclerosis (plaque buildup in the arteries) may be related to incidents that occurred prior to birth and that are not necessarily genetically determined. Interestingly, maternal malnutrition during the Dutch Hongerwinter was associated with a doubling in the prevalence of high-fat diets in their offspring at a later age.[21]

Half a century ago, the geneticist James Neel proposed a theory to shed light on the development of diabetes.[22] He coined the term "thrifty Genotype" as a process of gene selection by which humans who live in an environment where food is scarce develop the ability to use food as effectively as possible. Thirty years later, Charles Nicholas Hales and David J. Barker[23] expanded on this theory to present the "Thrifty Phenotype Hypothesis," which states that maternal signals conveyed to the fetus are instrumental for the development of a thrifty individual equipped for an environment which is scarce in food supply. The trouble arises with the mismatch of expectation and reality: if the environment is abundant with food resources, the capability to efficiently accumulate food turns into a disadvantage and these individuals become more likely as adults to develop diabetes and metabolic syndrome. Such individuals tend to be insulin resistant, which means that their body cells are unable to use insulin effectively and "resist" the function of insulin to pump glucose into the cells. As a result blood sugar increases.

If we look at the three factors involved: Insulin resistance, low birth weight, and increased risk for future disease, then the puzzle turns into a picture. Insulin is a growth factor and intrauterine insulin resistance leads to reduced birth weight. Insulin resistance is also a risk factor for future disease. So, reduced birth weight may be an indication of insulin resistance, rather than the other way around.[24] Interestingly, birth size and insulin resistance have been related to a specific receptor gene (PPAR-y2).[25] Individuals who were born preterm with very low birth weight tend also to engage less in physical activity as young adults[26] and to eat more later in life.[27] They also seem to prefer high calorie food, all of which predispose them to later obesity.[28] It has been shown, especially in women, that intrauterine growth retardation is related to a later preference of high-carbohydrate over high-protein diet and also to a larger waist/hip ratio. This was particularly evident in young women.[29]

Low birth weight, be it SGA (small for gestational age) or AGA (appropriate for gestational age) has been shown to be associated with later increased consumption of a high-fat diet but also with the reduced eating of fruits and vegetables. It seems obvious that this combination of highly unhealthy eating habits would lead to later disease. There are other complications as well. Mothers of babies born with low birth weight are usually happy if their baby's growth catches up quickly after birth.[85] Yet, growing evidence indicates that overly rapid increase in weight after birth is associated with diabetes later in life. It may be a cause for joy in the short term but is likely to be harmful in the long term.

FETAL PROGRAMMING— A LOOK TO THE FUTURE

Understanding the principle of fetal programming means recognizing the major impact of processes taking place during fetal life and their impact on the future health of the child. But that's not enough—to make an impact, we must push for major changes in the function of prenatal clinics and centers that care for pregnant women, infants, and children. Much more emphasis needs to be placed on the education of expectant mothers regarding the avoidance of toxins, proper nutrition, and nutritional supplements that can have a positive impact of the fetus. Pregnant women must be provided an environment at home and at work which takes into consideration the needs of their fetuses. These steps and others will expand the definition of prenatal care from care for pregnant women to a new model of medicine aimed at preventing diseases in adulthood. This is nothing less than defining prenatal care of the pregnant woman as preventative health care for humans throughout their life cycle. Medicine needs to make people aware of how much is at stake during the prenatal life of an individual relating to its future health and welfare. It is therefore imperative that appropriate resources be devoted to proper prenatal care. Peter Nathanielsz, the well-known British gynecologist and scientist, aptly summed this up[31]: "How we are ushered into life determines how we leave."

Let me offer a final comment to close this chapter. Fetal

programming should not be confused with determinism. An individual's fate is not sealed in its mother's womb. Education, awareness of possible risk factors, creation of appropriate treatment, and production of an environment appropriate to the newborn after birth serve to place much of our fate in our own (or our parents') hands in particular and in those of society in general. The understanding of fetal programming may be ultimately empowering—providing us with the tools to actively take control of our health.

CHAPTER 4

EMOTIONAL STRESS DURING PREGNANCY

D ESPITE ITS HIGH PREVALENCE, FEW OBSTETRICIANS ASK PREG-nant women about depression, anxiety, and stress in their lives. They worry rightfully about the mother's blood pressure or about gestational diabetes, but "maternal depression occurs at least as often as pregnancy-induced hypertension, and 5–10 times as often as gestational diabetes."[1] Modern obstetrics has not yet fully integrated the understanding of various aspects of fetal programming, including maternal stress on premature labor and delivery, for instance. Yet even minor stress on the mother affects the fetus. In an elegant experiment at Columbia University's Medical Center, pregnant women were asked to perform a stressful computer task while researchers monitored the fetal heart rate. A clear correlation was revealed between the degree of stress on the woman and fetal heart rate decelerations—that is, sporadically decreased heart rate.[2]

Stress is a response to stressors, to a perceived internal or external threat. Anxiety on the other hand is one of the possible reactions to a stressful situation, and includes anticipating misfortune with worry or distress. Certain stressors may cause different types and intensities of stress and anxiety depending on the person or the situation.

When we are challenged by stressors, our nervous system prepares us for a fight or flight reaction. Biochemical processes are initiated, our autonomous system is activated, and our organism recruits a variety of tools from many different bodily sources. Stress hormones like cortisol and adrenalin are secreted into the blood stream, our heart rate increases, our alertness is heightened and our body gears up to confront the perceived threat. At the same time certain functions and activities which are less important for our fight or flight response slow down, including our gastrointestinal system and even our immune system.

Given the interconnections we've covered so far, it seems obvious that the diverse physiological processes in a women's body under stress should affect also her fetus. Indeed, as early as 1972 a research study published in the prestigious journal *Science*[3] reported that pregnant rats under emotional stress during their pregnancies gave birth to male rats who exhibited female rather than male sexual behavior. Suppression of testosterone in male fetuses may lead to feminization, or feminine development of their sexual behavior. A connection between stress during pregnancy and cognitive disturbances in later life has also been reported in humans.[4, 5]

I should emphasize that most children born to mothers who were exposed to major emotional stress during their pregnancies will be perfectly healthy. An "increased risk" means just that, namely that a pre-existing risk may increase, but still leave the majority of children unaffected. Moreover, it is impossible, and probably not desirable, to avoid stress completely during pregnancy. A low to moderate amount of stress is even likely to be beneficial in the emotional and physical development of offspring. This finding emerged from studies on pregnant women who lived in an economically and emotionally stable environment and were exposed to modest levels of stress as evidenced by commonly used scale questionnaires.[6] The researchers from Johns Hopkins Medical Center in the USA reported in 2006 that mild to moderate emotional stress in healthy women during pregnancy contributed to improved motoric and cognitive development in their offspring as compared to the offspring of the women who were not exposed to similar stress events. With these caveats in mind, let us examine the growing body of data that links antenatal maternal stress with gender-specific impairment of emotional and cognitive development in childhood.

There is overwhelming evidence showing the detrimental effect of excessive stress in pregnant women on the brain and on the cognitive development of her fetus. More worrisome is that these effects, once expressed, continue throughout childhood, adolescence, and into adulthood. In rodents, it has been shown that prenatal stress causes epigenetic changes leading to a modification of brain receptors that bind the stress hormone cortisol. This mechanism has also been identified in women who were subject to domestic violence during their pregnancies and in adolescent children of these victimized women.[7]

And it's not only increased stress during pregnancy but anxiety or depression which may lead to an increased risk of impaired neurodevelopment, emotional and behavioral disorders, and to diseases associated with the immune system, such as asthma.[8]

Major emotional stress during pregnancy, for example in times of war or natural disasters such as floods, may cause irreversible damage to the neurological development of offspring throughout their lives. This may lead to shortened attention span, anxiety, defective cognitive functioning, and even increased incidence of schizophrenia. A study[9] published on a large group of 1.3 million births in Denmark between 1973 and 1995, gave the researchers access to information on the offspring for a period ranging from 10 to 32 years. During their pregnancies, some of the women received bad news regarding serious illness or the death of a first-order relative. The announcement of this bad news during the first trimester of pregnancy correlated with a significant increase in schizophrenia among the offspring, particularly in female children. Moreover, severe stress in the first trimester, such as the death of an older child, has been associated with an increase in congenital malformations of the newborn and premature delivery.[10]

The psychiatrist Jim van Os and colleagues[11] performed a study on pregnant women who were under severe stress during the May invasion of the Netherlands in 1940. The study demonstrated a 28% higher risk for schizophrenia in the children of women who experienced the invasion during the first semester of their pregnancy. In women who experienced it during their second trimester of pregnancy and whose offspring were later diagnosed with schizophrenia, boys had a 35% higher risk for schizophrenia than female girls. This finding implies that severe maternal stress during the first trimester may increase the risk for schizophrenia in both girls and boys but that this effect is extended into the second trimester in boys only. The authors attribute this fact to a slower pace of early cerebral development in boys and hence a higher window of vulnerability.

In another study, researchers examined patient files of close to 90,000 births which occurred in Jerusalem between 1964 and 1976.[12] This period included the Six Day War in 1967 and the Yom Kippur War in 1973, as well as the relatively quiet periods before and between the wars. Thus the researchers were able to compare pregnancies in emotionally stressful wartimes with those during periods of relative calm. The researchers compared their data to the Israeli psychiatric database in order to assess the relative risk of schizophrenia according to gender and other variables. The comparison revealed that the risk of developing schizophrenia was 2.3 times greater among individuals whose mothers were pregnant during times of war compared to those whose mothers were pregnant during peacetime. This study also found that this risk was four times higher among women than among men.

Natural disasters have a similar impact. Prenatal maternal stress as experienced by pregnant women during the North American Ice Storm of 1998 motivated large scale longitudinal studies in the framework of "Project Ice Storm."[13] Over 3 million people were without electricity for up to 40 days. One of the research projects followed children of about 150 women who were pregnant at the time of the storm and aimed to separate objective stressors (days without power) from the subjective reactions (post-traumatic stress symptoms). Such studies have provided significant evidence of behavior problems, impaired motor and physical development, as well as detrimental effects on IQ, attention, and language development in the offspring of affected women.[14]

The developing brain is sensitive to stress hormones, and changes in what are called endocrine axes can cause permanent changes in brain function. An endocrine axis is governed by certain regions in the brain that secrete hormones, which in turn stimulate target glands in the body, among them the thyroid gland, the testicles, the ovaries, and the adrenal gland. As a result of stimulation, these glands secrete other hormones that reach their target cells through the circulatory system, where they are bound to specific receptors. To close the circle, after target endocrine glands have produced desired levels of a hormone they "report" back to the brain regions that activated the original command chain, letting them know that the target gland no longer needs to be stimulated. This process is known as negative feedback (see Chapter 9). Thus, for example, increased secretion of cortisol from the adrenal gland in an individual under stress is appropriate as long as the normal negative feedback mechanism causes the cortisol level to drop when the perceived danger has passed.

But what if the negative feedback mechanism is broken? It has been reported[15] that emotional stress applied to pregnant rats caused irreversible damage to the offspring's hormonal system, including the axis between the brain and the adrenal gland (which among others secretes stress hormones). Consequently, the daily secretion pattern for these hormones became disturbed and there was an increase in the levels of stress hormones in the brain. This effect seemed to be time-sensitive, and impinged on the fetal brain only during specific time windows. Among males, testosterone levels were observed to drop to the levels usually found in females. Males were born with more learning disabilities, while among female rats more symptoms of anxiety and depression were observed. Prenatal stress in pregnant animals can cause permanent impairment in offspring, including shorter attention spans, anxiety, and impaired cognitive function.[16]

Likewise, in humans, mothers who suffer from increased anxiety during their pregnancies may be more likely to give birth to children with Attention Deficiency and Hyperactivity Disorder (ADHD). This condition is more prevalent among boys than among girls.[17] The negative effect of maternal stress on the fetus is mediated by maternal cortisol that enters the fetal system. Usually, maternal cortisol levels are over 10 times higher than those in cord blood but the fetus normally "protects" itself and takes an active role in maternal cortisol "management." The placenta normally releases a certain hormone (CRH) which is involved in the increase of maternal cortisol secretion and, at the same time, the placenta releases an enzyme (11β-HSD2) which inactivates maternal cortisol that reaches the placenta, thus protecting the fetus from being flooded with maternal cortisol.[18] This enzymatic activity is apparently sex dependent and is increased in females. However, once maternal cortisol levels increase to very high levels due to chronic stress, the "enzymatic dam" may become damaged and the fetal system is flooded with maternal cortisol.[18] This may affect the delicate feedback mechanism in the fetus and reduce the activity and the development of the fetal adrenal gland which under these circumstances secrete **less** cortisol. Hence, fetal blood levels of cortisol may not be altered or be altered only slightly, although the fetal system of stress response may already have been irreversibly damaged.

The pharmacological blockage of the placental enzyme in mice[19] leads to reduced birth weight, to high blood pressure, increased levels of blood sugar and anxiety behaviors in offspring. Reduced enzyme activity has also been associated with a variety of pathological conditions in pregnancy like high blood pressure in the mother and intrauterine growth restriction of the fetus.[20] This would be an example of a bi-directional effect of mother and fetus and how pathology in the placenta may affect maternal health.

• • •

High levels of anxiety and depression in pregnant women are also associated with adverse behavioral outcomes in the future life of their exposed fetuses. In a prospective longitudinal study it has been demonstrated that such infants are more irritable, more difficult to soothe, and cry excessively, compared to the control group. At age 9, these same children (again, more so for boys than girls) were more likely to show signs of ADHD and aggressive behavior.[21] On the other hand, ADHD may also have an inheritance aspect as revealed in a study on women who suffered stress during pregnancy as a result of in vitro fertilization.[22] One group of women were the biological mothers of their babies while the other group consisted of women who underwent IVF with egg donation. Children born after IVF with egg donation did not present an increased incidence of ADHD while those who were biologically related did. The authors concluded therefore that inheritance may play a role in stress-induced ADHD.

Anxiety and depression are both independent risk factors for a variety of obstetric complications such as operative delivery, premature birth, and low birth weight. The assumed mechanism behind these disorders is an abnormality in the fetal Hypothalamus-Pituitary-Adrenal endocrine axis.[23]

In addition to the hormone cortisol, two neurotransmitters called serotonin and dopamine may also be involved in programming the neurocognitive and behavioral development of children whose mothers were under prenatal stress. Serotonin acts also as a nutrient factor during pregnancy and is involved in cell division and the creation of brain synapses (connections between nerve cells). It has been demonstrated that increased serotonin levels may affect offspring behavior in animals and this could explain the detrimental effect of certain drugs known as selective serotonin uptake inhibitors (SSRI), such as Prozac, on the fetus. Recently, researchers have identified a placental source for serotonin which may be involved in fetal programming.[24] This is yet another example of how fetus-derived substances influence fetal programming.

Intrauterine stress is associated with other features that develop in utero, such as reduced telomere length, which by itself is associated with reduced life span.[25] Telomeres are the "caps" on the end of chromosome strands. They are thought to protect the chromosomes just as thimbles protect our fingertips. During repeated cell divisions, telomeres shorten—this shortening of the chromosome protectors is thought to be part of the aging process. Yet telomere shortening has been associated with smoking, obesity, poor diet, lack of exercise, and stress.[26] Since reduced telomere length is associated with reduced longevity, the old saying about stress and sorrow hastening death seems to be gaining scientific evidence.

• • •

The connection between body and psyche is one of the oldest problems in philosophy and fascinated scholars for millennia. Heraclitus, the Greek philosopher who lived approximately 2,600 years ago compared the psyche to a spider and the body to its web. Just as the spider is harmed by damage to its web and quickly runs to repair it, so is the soul of a human affected when part of the body is harmed and so does the body react when the psyche is stressed. Modern psychosomatics is based on the understanding that psyche and body are interconnected to a great extent and there is no doubt that stress, like anything else which affects the pregnant woman, may also affect her fetus. We need to keep this foremost in our minds when caring for pregnant women.

CHAPTER 5

GENDER ASPECTS OF
HEART DISEASE

T HE HEART IS AMAZING. ONE OF THE FIRST ORGANS TO DEVELOP in the body of the fetus, it begins working around three to four weeks after fertilization, when the fetus is only 5 mm long—no bigger than a pea. Yet, this tiny blood pump is sufficiently developed to begin its work and to continue without ceasing until the end of life. The inherent strength of this organ is tremendous. Over an average person's lifetime, his or her heart will beat more than three billion times and will transport around 120 million liters of blood. If the actions of the heart over an entire life could be reduced to one concentrated effort, it would generate enough energy to drive a truck to the moon and back.[1]

All the blood in our body passes through the heart approximately once a minute. Yet the heart, which serves the human body devotedly and tirelessly, does not benefit from the biblical injunction: "Thou shalt not muzzle the ox when he treadeth out the corn." (Deuteronomy 25:4) The heart cannot tap into the enormous resources of blood that passes through it for its own needs. In effect, the heart is completely dependent upon the supply of blood it receives from the blood vessels designated for this task (the coronary arteries). Any shrinking or narrowing of these blood vessels, a condition known as coronary disease, endangers the heart muscle.

Heart diseases have traditionally been considered male

diseases. More than half the men who had heart attacks before the age of 35 will not reach the age of 65.[2] Among men, diseases of the blood vessels usually appear 10 to 15 years earlier than among women, but women with risk factors can become ill with heart disease at a young age. So while there is truth to this view that heart disease predominantly affects men, it holds only until the fifth decade of life when women reach menopause. The main reason that heart diseases are less frequent among younger women is the protection given to the heart by female hormones, such as estrogen. After menopause, when the ovaries have stopped secreting estrogen and because of the increased prevalence of risk factors (such as obesity, high blood pressure, and high cholesterol) in middle age, heart disease rises sharply among women. Today, cardiovascular diseases—heart diseases, cerebrovascular accidents (CVA or stroke) and vascular diseases—are the top causes of mortality among women. In the United States, cardiovascular diseases kill more women annually—half a million women a year—than all forms of cancer combined. The hormonal system plays an important role in maintaining the health of the heart, and the hormonal changes that a woman undergoes during menopause have a major impact on her body in general, as well as on the reproductive system. It has been discovered that women who go through early menopause before the age of 40 live two years less than women who experience menopause later. Another study that monitored 5,000 women for a period of 22 to 59 years found that the mortality among those whose ovaries had been removed before age 45 and were not given hormone replacement therapy was 84% higher than among those whose ovaries had not been removed.[3]

Even after menopause, the ovaries continue to secrete sex hormones—not the female sex hormone estrogen, but male sex hormones like testosterone. This hormone plays a number of important roles. It slows down the loss of bone density and even helps to build new bone. Indeed, a direct correlation has been found between low levels of testosterone and the frequency of bone fractures (see Chapter 13). For women after menopause, the primary benefit of testosterone is that in different bodily tissues, primarily in fat tissues, it is converted into estrogen. Hormones from this group partly take over the protective role played by ovarian estrogen prior to menopause. Thus, the protective role played by female hormones continues throughout a woman's lifetime. Without it, there would be more cases of osteoporosis and higher risk of contracting Alzheimer's disease, cardiovascular diseases, and others.

Ovaries appear to have other protective functions as well. The results of a comprehensive study of 29,650 women who were monitored over the course of 25 years showed that oophorectomy (surgical removal of the ovaries) was associated with a 17% increase in the risk of coronary heart disease. This risk rose to 26% if the surgery was performed before the age of 45. Moreover, the risk of death from heart failure was 28% greater. Among women who underwent oophorectomy and did not receive hormone replacement therapy, the risk of stroke rose by 85% and the risk of coronary heart disease doubled.[4]

Doctors who advocate "preventative" oophorectomy when a woman undergoes surgery for removal of her uterus for non-cancerous reasons argue that it will prevent ovarian cancer in the future. This may be true, but in order to prevent one case of ovarian cancer, 300 women would have to undergo oophorectomy and would subsequently be at significantly greater risk of cardiovascular disease and osteoporosis. According to this calculation, if ten thousand women were to undergo oophorectomy around the time of menopause, 47 of them would avoid death due to ovarian cancer by age 80, but 838 more would die of cardiovascular diseases and 158 more would die of the complications of fractured hips.[4]

The ovaries are thus an important organ throughout a woman's life, even after menopause. While she is fertile, they play a role in reproduction and are responsible for maintaining the female hormonal environment. After menopause, they play no further role in reproduction, but their hormonal role continues, albeit to a lesser extent. It is important to know that even after the period of fertility, the ovaries continue to fulfill vital roles in a woman's body. Nonetheless, I should mention that estrogens are not effective in preventing heart disease if other background illnesses have already developed. In such cases, the use of estrogens may even make things worse.[5]

TOO LITTLE RESEARCH, TOO LITTLE TREATMENT: HEART DISEASES AMONG WOMEN

Leading cardiologists from five European countries attended a seminar held in Brussels, Belgium in 2010. The conclusions of their discussions were published in an article titled: "Red Alert for Women's Heart: The Urgent Need for More Research and Knowledge on Cardiovascular Disease in Women."[6] The authors of the article stated that, despite the significant differences between men and women in the presentation of most cardiovascular diseases, women are still not sufficiently represented in clinical experiments. They pointed to the rise in the prevalence of these diseases among women and the urgent need to study the basic biological difference between the sexes in this context. Women constitute fewer than 30% of the participants in clinical trials for cardiovascular diseases, and only half the scientific publications include analyses of the results according to participants' sex.[7]

THE ANATOMY OF THE HEART AND TYPES OF CARDIOVASCULAR DISEASES

In order to better understand how diseases occur in the heart, I'll briefly review its anatomy and physiology (Figure 1). The following brief description of the functioning of the heart is best understood while referring to this illustration.

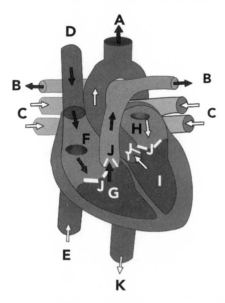

Figure 1. Anatomical structure of the heart.

A. Aorta, oxygenized blood to the upper body

B. Pulmonary Artery, non-oxygenized blood from the right heart to the lungs

C. Pulmonary vein, oxygenized blood from the lungs to the left heart

D. Superior Vena Cava, non-oxygenized blood from the upper body to the right heart

E. Inferior Vena Cava, non-oxygenized blood from the lower body to the right heart

F. Right atrium	**H.** Left atrium
G. Right ventricle	**I.** Left ventricle
J. Valves	

K. Aorta, oxygenized blood to the lower body

99

The heart's primary function is to transmit blood that has been oxygenated in the lungs to all the tissues of the body. The heart itself is composed of two parts, the right and the left, each of which comprises two cavities: the atrium and the ventricle. The atria receive blood through the veins, while the ventricles pump the blood out of the heart through the arteries. Oxygenated blood from the lungs enters the heart's left atrium (H) and passes into the left ventricle (I). As the ventricle contracts, the blood flows to the whole body through the arterial system. After the body has absorbed the oxygen carried by the red blood cells, the blood returns to the heart through the venous system. It enters the right atrium (F) and from there passes to the right ventricle (G). Contraction of this ventricle moves the blood to the lungs to be oxygenated again, and the process repeats. The passage of blood from the two large veins to the atria and from the ventricles to the two large arteries is timed to a precise rhythm, and the proper direction of the flow is maintained with the help of valves. Synchronization of the contraction and relaxation of the atria and ventricles and of the opening and closing of the valves is controlled by an electrical conduction system that includes, among other things, an internal pacemaker. The structure of this system is identical in the two sexes.

Having reviewed the anatomy of the heart, we can divide cardiovascular diseases into several types: those that affect the heart muscle through inflammation, diseases of the valves, diseases related to the electrical conduction system, diseases of the blood vessels supplying blood to the heart muscle itself, and coronary diseases that ultimately are likely to damage the heart muscle, leading to heart attack. As noted, the anatomy of the heart is identical in the two sexes, but it is interesting to discover that impaired heart function and changes due to disease can differ quite drastically between men and women.

RISK FACTORS FOR
CARDIOVASCULAR DISEASES

In this section I'll focus mainly on coronary heart diseases, mentioning other types of heart disease only briefly. Coronary heart diseases are diagnosed based on numerous risk factors, the most important of which include diabetes, high blood pressure, blood fat disorders, obesity, smoking, and lack of physical exercise. Risk factors more common among women include high blood pressure and diabetes, while among men smoking is the most prevalent risk factor. Certain diseases that appear during pregnancy, such as gestational hypertension, gestational diabetes, and abnormal prenatal development of the fetus are also considered risk factors for the development of cardiac disease and diabetes over the course of life. We will return to this point in a bit.

One of the major causes of cardiovascular (relating to the heart and blood vessels) disease and stroke (brain damage due to vessel disease) is a lack of oxygen in tissue. This may be due to the narrowing of the coronary arteries by the buildup of fatty sediments or plaques in the blood vessels. When this narrowing prevents sufficient blood flow to the heart muscle, the patient feels severe pressure on the area of his heart (angina pectoris), which can develop into a heart attack and damage the heart muscle. A blood clot can form on the sediments inside of the artery and at a given moment can lead to total blockage, causing a heart attack. If this happens in a blood vessel supplying the brain, the result would be a stroke. In men, heart attacks are usually due to a tear in the sediments on the inner layer of the artery, while in women it is more common to see a crumbling of the sediments rather than an actual tear. This process of tearing in men is referred to as explosion, while the crumbling process in women is known as erosion. These processes may happen anywhere in the arterial blood system from where particles of clots can be carried to critical organs the like heart or brain. In women erosion of the residue more frequently leads to the formation of smaller particles that pose a risk to the small blood vessels, including in the brain. This, perhaps, is the reason that strokes are more common among women, while heart attacks are more common among men. This gender difference in the prevalence of these diseases is probably one of the reasons that treatment with aspirin—a blood thinner—is more effective in preventing strokes among women and in preventing heart attacks among men. Blood thinners are used in order to prevent the building up of blood clots in small blood vessels.

Coronary heart disease is the most common cardiovascular illness that reduces the blood supply to the heart muscle. Current research points to a major difference between the way men and women are affected by coronary disease, which includes which populations are afflicted, the course of the disease, how it affects the body, and its prognosis. While we understand some of these differences, regarding a large part of them we are still groping in the dark. The rise in the number and severity of risk factors also increases the danger of illness and even death from heart disease.

The risk factors for coronary heart disease affect men and women similarly, but the **intensity** of their influence differs. For example, smoking under the age of fifty is more dangerous to women and raises their risk of heart attack even more than among men.[8] Moreover, because coronary heart disease appears at a later age among women, there is a greater risk that a woman will have other background diseases in addition, such as high blood pressure, obesity, and diabetes, factors that themselves constitute a risk for increased mortality.

Diabetes as a risk factor also has a greater impact on women, and women with diabetes develop heart disease more frequently than do men with diabetes. After menopause women sometimes gain weight, and this excess fat often accumulates around the waist, which increases the risk of metabolic syndrome, a cluster of conditions including high blood pressure, obesity, abnormal blood fat levels, and impaired sugar metabolism—a risk that is greater among women than among men. The tendency to develop diabetes also rises after menopause, and these factors together make the risk of fatal heart disease 50% greater among these women than among men.[9]

In addition, the prevalence of high blood pressure among women rises after menopause and by itself constitutes a risk factor for developing heart disease. Some of the symptoms of menopause, among them chest pains, rapid pulse rate, and even hot flashes, may also be related to increased blood pressure and not necessarily to menopausal symptoms.[10] After menopause the metabolism of fat changes in women and cholesterol levels begin to rise, including so-called "bad" (LDL) cholesterol. While the levels of "good" (HDL) cholesterol are usually higher in women, low levels of this class of cholesterol constitute more of a risk for heart disease for women than for men.

Coronary heart diseases such as arteriosclerosis (stiffening and narrowing of blood vessels) often develop differently in the two genders. Among men, the most common cause is fatty sedimentation or residue on the inner wall of the blood vessels. Images of cardiac arteries taken after inserting small tubes into the artery with contrast material reveal a typical picture of a blood vessel with a normal opening along its length but with various degrees of narrowing or blockage at one or more sites. Among a considerable number of women, in contrast, the images do not show localized narrowing but rather a thickening of the wall of the blood vessel along its entire length and a narrower opening along the entire passage. This is the case among 30% of women, primarily young women with coronary heart disease. Only 15% of men who have had heart attacks exhibit a similar picture. The mechanism of heart attack in the presence of non-blocked coronary arteries was enigmatic for years and was therefore named Cardiac Syndrome X. Today we know that the culprit in such cases is not the main cardiac arteries but rather the system of very small blood vessels and the inner lining of these vessels.

Therapeutic attempts to open the blockage with the help of a balloon or a stent inserted through a catheter or by open-heart surgery to bypass the blocked region are obviously less effective in these cases. Yet because treatment of heart attacks is usually geared to the "male" pathology of the heart, it is likely that the typical treatment to correct blockage of coronary arteries may not be suitable for women with heart attacks. Women need new therapeutic methods. This, among other reasons, is why gender medicine—and studying the path of coronary heart disease among female patients—is so urgent.

DISEASES OF THE HEART MUSCLE

As a rule, women are more likely to exhibit impaired functioning of the heart muscle during **relaxation** of the heart muscle, that is, after the heart has been emptied of its content, while men are more likely to exhibit impaired heart muscle functioning during **contraction** of the heart.

Disordered changes to the heart muscle differ between men and women as well. Among men, in response to the heart's efforts to operate under conditions of ongoing high blood pressure, the heart's left ventricle, which serves as a pump to transfer oxygenated blood to the body, expands and its walls become thinner, resulting in a kind of ballooning. Among women under similar conditions the heart does not increase in size but its walls become thicker, reducing the heart's capacity and increasing the risk of stroke. This process seems to be caused by the hardening left ventricle's increasing dependency on the proper functioning of the left atrium, which supplies it with blood. Unfortunately, among women it is more common to find that the left atrium is incapable of properly filling the left ventricle, in particular in a state of fibrillation—that is, very rapid and ineffective contractions. Thus, the left ventricle, which has shrunk, receives an even smaller supply of blood to distribute to the various regions of the body.

Atrial fibrillation is extremely dangerous. In addition to reduced blood flow to the body, including the brain, the greater coagulability of the blood in the atrium that is not contracting efficiently poses a risk that small clots will form and be carried to the brain, where they could cause a stroke. This is why the risk of stroke as a result of atrial fibrillation is greater in women than in men.

GENDER ASPECTS OF
HEART DISEASE DIAGNOSIS

The differences between men and women in the functioning of healthy and diseased hearts suggest that diagnostic methods are not equally suitable for both genders and that the symptoms of the same disease may vary between individuals—more than we had previously believed. Gender medicine is making it clear that different methods of diagnosis and treatment are necessary for men and women.

Among men, the cardiac stress test that indirectly measures blood flow to the heart muscle has high specificity and is quite effective. That is, if the results point to a problem, in most cases, a problem indeed exists. Among women its specificity is lower, meaning that indications of illness may appear when there is in fact no problem. Consequently, many women will undergo unnecessary interventions. Moreover, the stress test is less sensitive among women, meaning that coronary heart disease is often not discovered by the test. Thus, the effectiveness of cardiac stress tests and of single-photon emission computed tomography imaging (SPECT) is limited in women. Results of SPECT testing can be erroneous and yield false positive results —that is, they can indicate disease that does not exist. This happens particularly in cases of women with large breasts, diminished heart size, or smaller cardiac blood vessels. In order to properly evaluate the results of stress tests in women, other tests are necessary, yet these are not routinely done.[11] The preferable test for evaluating heart disease in women is the stress echocardiogram, which uses ultrasound imaging to show how well the heart muscle is working to pump blood to the body.

Thus despite such disparities, most research studies continue to be conducted on men only. The accumulated knowledge about heart health in men is not necessarily applicable to women, making it ever more difficult to diagnose and treat women's heart problems effectively.

HEART ATTACKS IN MEN AND WOMEN

The classic picture of a heart attack is well known. The textbooks usually describe a slightly overweight man grasping his left chest with his right hand after experiencing sudden chest pain. The pain radiates to his left shoulder and arm, and the

106

man's facial expression shows that he is scared to death. Not only will a first-year medical student immediately conclude that he is having a heart attack, so will most of the general public.

The problem is that among 20% of women, heart attacks look completely different. The symptoms do not appear suddenly but may develop over a number of hours and even days. The woman may suffer from shortness of breath and the "typical" pain may seem quite atypical, radiating to the back of the neck or the jaw instead of to the left shoulder. Women of all ages having heart attacks report much less chest pain than do men.[12] Women also complain more of general symptoms like nausea rather than of specific, localized symptoms.[13] A woman having a heart attack but without its classic symptoms will come to the ER much later than a man will, and the chances that she will be sent home without being diagnosed are two to four times higher than those of a man experiencing chest pain.

Because immediate medical intervention is critical prior to and during a heart attack, a delay in diagnosis can be fatal. An acute heart attack is more prevalent among men as the first sign of coronary disease. Nevertheless, more women die of this condition before reaching the hospital or after being released from the emergency room with the wrong diagnosis. The data is distressing. A study published 35 years ago which examined the rate of proper diagnosis of heart attack among 390 patients found that twice as many women received an incorrect diagnosis. Even when imaging tests pointed to problems, ten times more men than women were referred for further examination. Even worse, these incorrect diagnoses were not only due to differing symptoms among men and women. The same unequal treatment was evident even when actors from both genders presented with identical complaints.[14]

That was the situation 35 years ago. What about today? We have made some progress, but not nearly enough. Many doctors still have not internalized the notion that cardiovascular diseases aren't restricted to men only. A study published around eight years ago in England examined diagnostic and treatment methods among 1200 men and women with severe clinical angina pectoris (chest pain caused by suddenly reduced blood flow to the heart muscle). The results of this study are worrisome. The treatment women received was inferior to that given to men. Moreover, among women, cardiac risk factors were less documented, they were given fewer preventative measures and fewer referrals for cardiac consultations, and they underwent fewer angiograms.[15] Recently, a large study of 65,000 patients with acute coronary syndromes from 99 medical centers in France revealed that significantly more women than men had open coronary vessels and fewer women than men underwent coronary intervention in the form of insertion of stents or bypass surgery.[16]

HEART FAILURE

The prevalence of heart failure among the adults in developed countries at a given time is between 1% and 2% and is more common in men than in women. This is because the most common cause, namely coronary heart disease, occurs earlier in men than in women. With advancing age, the prevalence becomes equal between the sexes. Heart failure is relatively common among people over the age of 70, reaching a prevalence of over 10% within this age group.[17] With increasing life expectancy, the prevalence of this condition is expected to rise even more.

The causes of heart failure differ for men and women. Among men the cause is usually coronary heart disease or past heart attack, while among women the condition is typically related to background illnesses such as high blood pressure and diabetes. In both, life expectancy after diagnosis and mortality rate for this condition are similar to that for cancer.

Yet, there are gender differences. A multi-center study in Israel comprising 2,200 patients hospitalized for heart failure examined the impact of gender on mortality rate. Of the patients examined, 45% of them women and 55% men, a significantly greater proportion of the women had high blood pressure. The women were also older and there were fewer cases of coronary heart disease. The results of this study show that women were at greater risk of dying within six months after hospitalization while men were at significantly greater risk dying more than six months after hospitalization.[18] These results do not contradict the well-known paradox, according to which women with heart failure survive longer and in a better condition than men despite not receiving similar-quality care. They might indicate that women are at a more immediate risk of mortality after heart failure than men, but once recovered they have a better chance for survival than men. The situation is apparently different when patients with heart failure suffer a sudden cardiac arrest. Researchers from France[19] analyzed data from 13 studies including over 400,000 patients and reported that the survival rate among women was better than among men after discharge from the hospital. The authors explained this outcome by a better response of female heart cells to lack of oxygen and a stronger response of the their nervous system.

MEDICATIONS

Like the diseases themselves, the medications given to treat cardiovascular diseases were for the most part tested on men only. Yet some of these medications operate differently on women. Some are less effective, some have more side effects, and some endangered women to the point that pharmaceutical companies were forced to remove them from the market. For example, medications from the family of angiotensin converting enzyme (ACE) inhibitors (an enzyme that acts to maintain blood pressure) are less effective in women and cause more side effects. As mentioned earlier, aspirin administered as preventative treatment is more effective among women in preventing strokes, while in men it is more effective in preventing heart attacks. Another example is beta-blocker drugs intended to block the effects of adrenal hormones on the heart and to treat cardiac arrhythmias (irregularities of heartbeat) and high blood pressure. It seems that these drugs are likely to cause a particular and dangerous form of tachycardia (very rapid heartbeat) in women known as "torsade de pointes," which can lead to death.[20]

Ongoing emotional stress is also a risk factor for developing heart disease. In this case as well, the impact is greater among women than among men. An extreme form of how emotional stress differentially affects women is illustrated in the following description: There are certain heart diseases that appear almost exclusively among women, such as Takotsubo Cardiomyopathy or "broken heart syndrome," a condition that appears after menopause and imitates the classic symptoms of heart attack. ECG tests reveal changes similar to those that occur with heart attacks and blood tests show elevated levels of certain enzymes typically secreted during heart attacks, but imaging studies show no blockage of the cardiac blood vessels. Nonetheless, even though heart function drops sharply in this condition, it is not a case of heart attack but rather a form of temporary exhaustion and cardiac failure caused by a sudden flooding of stress hormones such as adrenalin. These hormones are generally secreted as a response to sudden and severe emotional stress, like the news of the death of a loved one, a natural or manmade disaster, or sudden and acute physical pressure. In imaging studies, the heart appears fatigued, the apex of the heart and the adjacent regions do not contract properly and the heart resembles the shape of fishing pots used by the Japanese for trapping octopus (*tako* = octopus; *tsubo* = pot). This disease does appear in men, albeit much more rarely. Men with this disease are usually younger and in men the preceding stress is more often physical while in women it is usually emotional. Moreover, cardiac complication during hospitalization, including cardiac arrest, is more common in men than in women.[21] In its acute stage, this condition may be life-threatening, but the damage to the heart is reversible. Because it is not a heart attack, treatment for this condition must include calming the patient down and offering emotional assistance and empathy. In almost all such cases, the heart returns to normal within a short period of time. To be on the safe side, most cardiologists will recommend ongoing preventative treatment. The "broken heart" syndrome is yet another example of the truth behind the folk wisdom that described heartache as a result of sorrow long before its scientific basis was discovered.

THE HEART OF A WOMAN: A LOOK AT THE FUTURE

Pregnancy offers a unique opportunity for predicting the development of heart disease in women. For example, gestational hypertension is a cluster of conditions occurring during pregnancy that may include increased blood pressure, increased secretion of proteins in the urine, delayed growth of the fetus, and more. Some of these conditions require intensive treatment during pregnancy, while most of them pass after delivery. Yet women who suffer from hypertension during pregnancy are at twice the risk of developing hypertension and cardiovascular diseases later in life than women who do not. The same is true for gestational diabetes. This condition usually passes after delivery, yet it increases a woman's risk of developing diabetes later in life and is additionally a significant risk factor for developing heart disease. Placental insufficiency—which can lead to a delay in fetal growth or intrauterine death of the fetus—is also a warning signal that a woman may develop cardiovascular disease after giving birth.[22]

A series of studies conducted in southern Israel that monitored births from high-risk pregnancies for a period of more than ten years found that the rate of cardiovascular illness among 5,000 women with gestational diabetes was almost three times greater than among women whose pregnancies had no complications.[23] The study also found that of 6,000 women who delivered prematurely, the risk of cardiovascular illness in the future was 50% higher compared to those who delivered normally.[24] Gestational hypertension also raised the risk of cardiovascular disease. The rate of future hospitalizations due to cardiac problems among 2,000 women with gestational hypertension was twice as high and the rate of chronic hypertension was 16 times greater.[25]

It is therefore hard to understand why an in-depth obstetric case history is still not part of cardiologists' guidelines in diagnosing and treating heart disease in women. Leading medical centers worldwide are now introducing gender-oriented units for studying, diagnosing and treating women's hearts, focusing on, among others things, women who had gestational conditions predicting a future risk of cardiovascular disease. These units take advantage of the window of opportunity provided by pregnancy to predict risk factors for cardiac and other diseases.

In summary, despite advances in medicine over the years, scientific knowledge regarding cardiovascular disease among women still lags far behind. For this reason, women with heart disease may still be subject to incomplete diagnosis and inferior treatment. If we want to keep women healthy, it is urgent that we include more women in medical research and teach and emphasize this topic in medical schooling and training.

CHAPTER 6

THE DIGESTIVE TRACT— GENDER ASPECTS

M OST OF THE ACTIVITY OF THE DIGESTIVE TRACT OCCURS spontaneously, with little control from the brain or the central nervous system. So long as the digestive tract works properly and we have no problems with digestion or with the frequency and consistency of bowel movements, we choose not to think about it at all. However, when the digestive tract or one of its components fails to work normally, suddenly we cannot help but focus all our attention on it. Romance disappears when we have a stomachache, the muses abandon us when we are suffering through a bout of diarrhea and heartburn blocks and prevents any hope of creative thinking.

In fact, functional disorders of the digestive tract are quite common. Constipation, diarrhea, heartburn, stomachaches, and excess digestive gases are all unpleasant occurrences and affect one in five people in the Western world. A substantial number of visits to the family doctor are due to digestive problems, and around 25% of visits to gastroenterologists are related to such disorders. In most cases, the causes of these problems and the mechanisms governing them are still a mystery to us. Laboratory testing generally does not uncover the origin of the problem and most of the time there is no anatomical explanation.

From the perspective of gender, functional digestive disorders affect men and women in different ratios—though it seems to differ by geography. For example, in western societies, irritable bowel syndrome (IBS), an extreme expression of these functional disorders, is four times more common among women than among men,[1] while in countries in East Asia such as Japan, China and India, the ratio between the sexes is the opposite. This finding indicates that this syndrome is not necessarily related to a biological difference between the sexes but rather to environmental differences, including differences in our microbiota, that is, the population of microbes in the large intestine (more on this in the next chapter). Taking into consideration that in the Western World, IBS is mostly prevalent among women, that for the most part it emerges between their late twenties and their mid-forties, and that it worsens during and immediately after the menstrual period, we can infer that hormones play a role.

Irritable bowel syndrome is usually classified into three types: diarrhea-predominant (IBS-D) which is more likely to occur in men, constipation-predominant (IBS-C), which is more likely to occur in women, and alternating stool pattern (IBS-A). Some research on this subject has shown that medications used to treat IBS have a differential impact on men and women.[2] Studies have also found a clear emotional factor in functional bowel disorders. Around half of those diagnosed with such disorders also suffer from emotional stress or mental illnesses.[3] An interesting example can be seen in a study conducted in southern Israel focusing on increased complaints of irritable bowel in the Bedouin population. This increase in the incidence of IBS was observed among Bedouins who moved from their traditional residential areas to permanent towns, the dramatic change in lifestyle causing significant stress.[4]

Before we can discuss gender aspects of the digestive system in more depth, however, we must first review how a healthy system works.

A TOUR OF THE DIGESTIVE TRACT

Ostensibly, the function of the digestive tract is simple: to absorb food, to break it down into small portions that can be transported as the "fuel" to maintain and operate the bodily systems, and to provide the raw materials for repairing and building cells in the body. After the food has been absorbed, the digestive tract is also responsible for removing wastes.

This process sounds simple, but the activities "break down," "prepare," and "remove" are supported by a powerful, sophisticated system. To describe its complexity in brief, we must add to this short list a number of essential functions. The wall of the intestine serves as a defensive barrier that protects the "inside" of the body. By inside I refer here to the contents of our body between the skin and the intestinal wall. Like skin, the wall of the intestine separates the outside world from the inside of our bodies. Technically speaking, we are hollow creatures, and the digestive channel that runs through our bodies is in effect part of the outside world. We eat and swallow, yet every bite we take can return to the place from where it came. Components of our food, such as certain fibers, exit the digestive channel at the other end, but except for having been chewed, they have not undergone any substantial change. Only food that crosses the wall of the intestine actually enters our bodies. How the food enters—in what form and what size—is controlled by the digestive system.

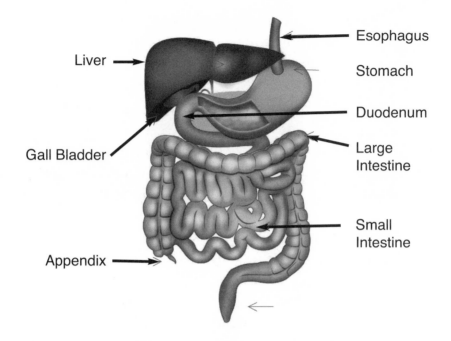

Figure 1: The Digestive tract

So what stations does our food pass through in its journey through our digestive system? Here I'll provide a brief overview while focusing on gender differences.

At the first station, the mouth, chewing and saliva break down food into small portions and enable the tongue and the nose to determine its quality using the senses of taste and smell. Saliva preserves the moisture of the mouth and throat, wets food while it is being chewed and helps break it down to prepare it to move to the next station. (More about saliva in Chapter 9). Even at this early stage significant gender differences are obvious. For example, jaw movements in male rats differ compared to female rats, even when they are still in the uterus. In humans, women secrete less saliva over their lifetimes, the composition of their saliva is different, and during their years of fertility, this composition differs throughout the menstrual cycle.

Chewed food moves from the mouth through the esophagus, a food pipe leading from the mouth to the stomach, and passes through a ring-shaped muscle called the lower esophageal sphincter before entering the stomach. Food's entrance into the stomach as well as its exit following digestion occurs according to precise timing. This is important because the stomach juices contain a high concentration of hydrochloric acid, which is strong enough that a small amount is able to perforate a piece of cloth. (Our stomach lining is protected from a similar fate, thanks to an impenetrable layer of mucus and cells that cover the wall of the stomach and continually renew themselves.) If the lower esophageal sphincter does not close hermetically and the stomach juices splash the wall of the esophagus even slightly, we will experience heartburn and in extreme cases, actual damage to the esophagus. The esophagus is shorter in women than in men and its contractions while swallowing are stronger to the point that they sometimes resemble the chest pain of a heart attack. Disorders in the function of the muscle between the esophagus and the stomach and heartburn are more common among men.

Swallowing is a conscious process under our control, that is, under the control of our brain. However, from this point on and almost until the food matter reaches its exit—the anus—the brain passes control over to the involuntary nervous system of the digestive tract, which has almost complete autonomy.

At the next station, the stomach, the food undergoes a coarse process of digestion, though not a complete process of absorption. Hence, the food is not yet *inside* our body but rather in the digestive channel. In a certain sense the food is still outside our body. Here we come to another fundamental gender difference in the functioning of the stomach. Emptying out the stomach, and indeed emptying out the entire digestive channel, is much slower in women, and the makeup of the digestive juices is different between the two sexes. For example, in men the enzyme responsible for breaking down alcohol is five times more concentrated than in women. While this enzyme is located mainly in the liver, it is also found in the stomach. Because of its relatively low concentration in women, the female digestive system breaks down alcohol less efficiently, which is why women become inebriated more quickly than men from the same amount of alcohol. According to a recent report, mortality due to heavy consumption of alcohol is significantly higher among women.[5]

Entry to the small intestine is safeguarded by a very alert "gatekeeper" that facilitates the synchronized opening of the stomach and passage of food in the exact quantities required for continued digestion. In other words, the stomach serves as a storeroom for food as well as a processing plant. In some cases this gatekeeper muscle operates too slowly or too late or becomes paralyzed. Such problems are usually caused by nerve damage as a complication of diabetes or occurs following injury to the vagus nerve from surgery. This nerve, while not essential to our body's function, is nonetheless important for inducing the intestinal contractions that cause movement through our digestive system. Such conditions, too, are most common among women.[6] It takes twice the amount of time for food to pass through the small intestine in women as it does in men. (Which raises the question: is it logical that men and women eat breakfast, lunch, and dinner at the same times? Would it be healthier if women ate twice per day and men four times per day?) As we will see in a later chapter, not only food, but also medications are retained longer in the female gastrointestinal tract, which has implications for drug absorption.

The speed of passage is regulated, among other things, by estrogen. After menopause, food passes more quickly, and during pregnancy, when estrogen levels rise, the passage speed decreases.

From the stomach, food passes to the duodenum, the upper part of the small intestine. There, and all along the small intestine, delicate processes of breaking down the food occur, much like separating wheat from chaff. The small intestine's most important function is to absorb food components through its wall into the circulatory and lymph systems. Digestion in the small intestine requires enzymes, which act as catalysts for breaking down food—particularly fats. Enzymes originate in the pancreas and in the gallbladder, which collects and concentrates the liver's waste products.

The drainage systems from the pancreas and the gallbladder that empty into the duodenum can become blocked, a potentially life-threatening condition. The typical patient with gallstones is an obese woman in her forties who is fertile and has a family history of gallstones. (In fact, medical students are taught to memorize the five F's that mark the classical type of an individual likely to suffer from gallstones: Fat, Female, Forty, Fertile, Family.) Here as well, female biology makes women more vulnerable to digestive malfunction. The gallbladders of women contract more slowly, bile passes more slowly, and its composition is different from the bile of men. Gallstones are four times more prevalent in women than in men, in particular when the level of estrogen rises as it does in pregnancy.

Sometimes the strong digestive juices from the stomach find their way to the duodenum. In such cases, even small amounts are sufficient to damage the intestinal mucous membranes and potentially develop ulcers. In this case, men are at a higher risk for problems. Such ulcers are twice as common among men than among women. Pre-menopausal women appear to be relatively protected against ulcers due to the estrogen in their bodies. Yet, ulcerative inflammatory diseases of the intestine, such as Crohn's disease and colitis, are 1.5 times more common in women than in men. The reason for this may be environmental rather than biological: in east Asia these diseases are more prevalent among men.

The next station is the large intestine or colon. One of the major functions of this organ is to preserve liquids and protect the body from dehydration. Between nine and ten liters of water pass from the small intestine to the large intestine each day, and under normal circumstances 99% of these liquids are reabsorbed there. Thus, the liquid that enters the large intestine transforms into solid feces which are expelled at given times. The large intestine is also home to our gut microbiota, trillions of microbes with whom we live in symbiosis, and without whom we could not survive. I'll go into more depth on this in the next chapter, which is devoted to the large intestine.

GENDER ASPECTS OF DISEASES
OF THE LARGE INTESTINE

Cancer of the colon is considered a silent killer, since symptoms appear relatively late in the course of the disease. Around one million cases of cancer of the large intestine and the rectum (colorectal cancer) are diagnosed worldwide each year. Half a million people across the globe die annually of this disease.

Only a few factors are known to increase the risk of such cancers (although most people who develop colorectal cancer have none of these). Among these are older age, the presence of polyps in the large intestine, a family history of colorectal cancer, a high cholesterol and low fiber diet, diabetes, lifestyle factors such as smoking, alcohol, and little exercise and more. As with other diseases, colorectal cancer affects men and women differently. In general, colon polyps, which are thought to be pre-cancerous growths, as well as colon cancer, are more prevalent among men. The incidence of this morbidity among women reaches that of men four to eight years later. There are other observations of sex-related differences in diseases of the colon, for which we have no explanations yet: Colon cancer is more common on the right side among women and on the left side among men. Women who smoke are at higher risk of developing colon cancer than male smokers. Women more often develop cancer of the large intestine and men more often develop cancer of the rectum. Women who have had cancer of their ovaries, uterus, or breasts are also at an increased risk of developing colorectal cancer.

Perhaps the difference in incidence of such diseases has to do with the gender imbalance inherent in diagnosis. Unfortunately, sex differences in the functioning of the gastrointestinal system are still not sufficiently appreciated by the medical system and this extends also to diagnosis: The following is an example of how diagnostic tools, which were developed mainly for men, may put women at a dangerous disadvantage: The most commonly used diagnostic tool for early detection is the examination of subtle traces of blood in the stool. It appears that fecal occult blood testing, as it's called, for early detection of colon cancer is less sensitive in women than in men. The reason for this is possibly related to the fact that the content of the colon moves more slowly in women than in men. Thus small amounts of blood which would be indicative of pathology in the colon, including cancer, also remain longer in the large intestine before being expelled. Yet, the longer the blood remains in the digestive system, the more time the red blood cells have to become oxygenated, and therefore liable not to be detected by the standard test. This would lead to a delay in diagnosis of colon cancer and since the time period between diagnosis and treatment is crucial for treatment success, women may pay with their lives.

CHAPTER 7

OUR INTESTINES—THE SECOND BRAIN AND LIFE WITHIN THE GUT

O UR LARGE INTESTINE IS HOME TO AROUND A THOUSAND different types of microbes that weigh close to two kilograms. Not merely a factory for removing waste and absorbing liquids, our digestive tract also includes a sophisticated nervous system which resembles in many ways our central nervous system. In this chapter we will explore the complex workings of our intestinal nervous system and gut microbiome in hopes of understanding why women and men suffer from different digestive problems—and to take the first step in improving treatment for both sexes.

THE SECOND BRAIN

The control of the digestive system is unique in its relative lack of reliance on the brain. The complex and sophisticated functioning of the stations along the digestive tract as described in the previous chapter naturally require precise neurological supervision as well as well-coordinated feedback. Yet it seems the brain—which otherwise commands the functions in the rest of our body—has given the gut a large degree of functional autonomy. A great deal more information goes from the digestive system to the brain than in the opposite direction. Only a few thousand nerves connect the brain and the hundred million nerve cells located in the digestive system, indicating that central control is of secondary importance as far as the brain and intestines are concerned. This turns out to be a brilliant adaptation. Central control of complicated peripheral systems is ineffective in most fields—in economics, in the military and in politics—and it's certainly the case in physiology. Highly complex systems such as the digestive system must operate with some degree of autonomy. To this end, the digestive system relies on an independent nerve system which Michael Gershon, a renowned American medical researcher, has dubbed "the second brain."[1] This is also the title of his groundbreaking book, and I rely upon some of his research in this chapter.

The study of the autonomic nervous system of the gut is known as neurogastroenterology and was established more than one hundred years ago when two British scientists, Ernest Starling and William Bayliss, described what they termed the "law of the gut."[2] Starling and Bayliss discovered that a rise in pressure in the intestines causes wavelike contractions (peristalsis) in the direction of the anus. To their surprise, they found that an isolated intestinal loop, completely disconnected from the body and placed inside a liquid-filled container in the laboratory, responded in a similar manner. Thus the researchers discovered that the wall of the intestine contains an independent nervous system. Today the law of the gut is known as the "peristaltic reflex." Yet the term "reflex" here differs from other reflexes in the body. Muscle reflexes or pupil reflexes are controlled by the brain, and if the neurological connection is severed, the reflex disappears. Only the peristaltic reflex of the intestine continues to operate without any connection to the brain and even in a state of brain death. (The heart, too, is able to continue functioning without being controlled by the brain. But continuous functioning in the heart is not related to a reflex action, but to an independent pacemaker, which is located within the heart muscle.)

Following Starling and Bayliss's discovery, German researchers later described the clusters of nerve cells between the layers of muscle in the intestinal wall and beneath the intestinal mucous membranes. Today we know that a complex autonomic nervous system is situated all along the length of the intestines. This intestinal nervous system, or second brain, is unlike anything anywhere else in the body except for the central nervous system. It comprises more than one hundred million nerve cells, more than exist in the spinal cord. In effect, it's a huge system for processing data and carrying out actions without any brain intervention.

The brain supervises bodily functions with the help of two dedicated systems. The voluntary nervous system is so named because it is used to operate all the muscles in the body as we will it. We decide whether and when to move one of our limbs, for instance. The nerve fibers reach the target organ directly from the brain, without any intermediary stations. The command to move the big toe originates in nerve cells in the brain and the conveying nerve fibers travel a long path. The second system is autonomic and involuntary, for example the heart muscle which operates without our consciousness. The nerve cells of the involuntary autonomous nervous system are located in the brain and in the spinal column. The operation of some organs can be both voluntary and involuntary, like the diaphragm, which increases and decreases the size of the chest cavity and thus causes the lungs to expand and contract, enabling us to breathe. We can control the depth of our breathing and its rate, but we also breathe unconsciously, most notably while we sleep. The autonomous nervous system is unique in that the fibers of its nerve cells do not directly reach the target organs.

There is always at least one additional nerve cell along the path to which the primary nerve fiber conveys the impulse—that is what regulates the muscle stimulus.

The secondary nerve cells are located in the spinal column or in the target organ itself. The junction between two neurons is called a synapse, and information is transmitted from one neuron to another with the help of special chemical substances known as neurotransmitters. When information transmitted from the neuron to the target cell passes through an intermediary station, its command can be adjusted or even changed. Therefore, the autonomic nervous process is more complex and subtle than the voluntary process and is subject to changes dictated by the environment. The autonomic nervous systems are controlled by the central nervous system. Yet for the most part the intestinal nervous system is not subject to this control and can operate independently; it is often referred to as a third autonomic nervous system. Still, the brain did not completely give up its control—the autonomic vagus nerve that originates in the brain plays a role in operating the digestive system starting with the esophagus and along most of the length of the large intestine. What is interesting is that the digestive system can function without this nerve. Moreover, emotional states have an influence on the functioning of the digestive system. Statements such as "I felt a lump in my throat," "my stomach turned over," and "I was so scared I got diarrhea" are not mere exaggerations, but describe the influence of the brain on the digestive system.

The central nervous system has retained control over two functions of the digestive system, however: swallowing and excretion. For obvious reasons these two processes require a voluntary, cognitive component. Without control over swallowing, we would not be able to plan our eating and drinking, and without control over excretion, we would not be able to maintain a social life. However, practically all the other actions of the gut can occur without any brain involvement and even after brain death.

Hormones influence the workings of the digestive system, and the primary hormone involved in this process is serotonin. This hormone is produced in the brain, in blood platelets, but mainly in the digestive system, and is connected to a wide variety of conditions. Serotonin transmits messages from one nerve cell to another and is associated with learning processes, sleep, blood vessel contraction, appetite, migraines, and even maternal feelings. Low levels of serotonin are related to depression, social isolation, and aggression. Moreover, gut serotonin plays an import role in gastrointestinal motility and serotonin levels in the general circulation have been linked to various diseases, such as irritable bowel syndrome, osteoporosis and even heart disease. It is interesting that 95% of serotonin is found in the digestive system and researchers from the USA have shown recently that gut bacteria regulate the local production of serotonin.[3] Here is a fascinating connection between brain function, mood, gut microbes, function of the gastrointestinal tract and gender aspects: If one considers that women have lower levels of serotonin in their bloodstream and that these levels also change in the course of the menstrual cycle, then one may find here a possible explanation for specific differences in many diseases and malfunctioning bodily systems, including the digestive tract.

• • •

The nervous systems of the brain and of the gut are very similar. Therefore, it should not come as a surprise if disorders of these systems are similar as well. Despite the autonomy of the digestive system, the brain remains the commander in chief and maintains an influence on digestive function. The concept of an anal personality and other psychiatric conditions related to the digestive system are well known even to non-medical professionals. Yet the more intriguing question is the inverse: whether the digestive system, based upon its neurological autonomy, can develop "neuroses" that originate in the gut. Can the second brain respond independently to physical or emotional injury? If so, in certain cases of emotional injury should the proper treatment be aimed directly at the digestive system?

One amazing discovery demonstrates the similarity between the "first brain" and the "second brain" in the manifestation of certain diseases such as Alzheimer's and Parkinson's. Changes identical to those in the brain have been found in the neurons of the gut, even before symptoms appear in the central nervous system! In *The Second Brain*, Gershon suggested that perhaps in the future it would be possible to diagnose damage to the brain by means of sampling intestinal tissue. In fact, this future has already arrived. Early cases of Parkinson's disease can be diagnosed today by intestinal biopsy.[4] Parkinson's disease is characterized by deposits of a certain protein called Alpha-syn in brain cells that normally produce a neurotransmitter called dopamine. The continuous loss of these cells leads eventually to a critical decrease in local dopamine levels and to the symptoms of Parkinson's. Researchers have found that deposits of the protein in nerve cells outside the brain and particularly in the gut may precede damage to the brain by years. Thus, simple biopsies of the large intestine could detect early-stage Parkinson's disease—a truly amazing achievement which may open possible venues for treatment.

Another disease contributing to our understanding of the relationship between the autonomic nervous and the digestive system is a congenital condition known as Hirschsprung's disease. This hereditary disease occurs in about one in 5,000 of live births and affects around four times more boys than girls. In Hirschsprung's, the lower part of the colon lacks nerve cells along the entire intestinal wall, effectively cutting it off from all nervous activity. Even the brain cannot activate the peristaltic waves to move substances through the intestines; although the intestinal channel is open and the central system is operating properly, nothing can move through it. Without treatment, which involves cutting out the damaged region, this disease is fatal. Yet through the closer study of this disease we may gain a deeper understanding of the nervous control of the digestive tract.

OUR GUT MICROBIOTA

As mentioned above, our large intestines are home to a huge and complex variety of microbes with which we live in a symbiotic relationship. Collectively they are called the gut microbiota. In a recently published book, researchers from the Weizmann Institute in Israel have come to the conclusion that approximately 40 trillion microbes live in and on our body—two times more than the number of body cells.[5] Moreover, the number of microbial genes is 100–150 times larger than that of human cells in our body. These microbes are essential to our health and most are found in the digestive tract. Some produce vitamins, like vitamin K, and some are involved in breaking down carbohydrates that were not digested in the small intestine. Some of the "friendly" microbes play a role in absorbing fatty acids such as "bad" cholesterol (LDL) and others support the immune system. Some protect us from dangerous microbes living in the large intestine. The gut microbiota is also involved in what has been termed the brain–gut–microbiota axis, constituting the complex interaction between the brain and the intestinal system which affects not only the function of our digestive system but our cognitive processes as well.

Modern technology has enabled us to classify these bacterial populations into five subgroups, and the size of these subgroups, both absolute and relative to each other, has been related to sickness. The group Proteobacteria increases in size during disease and is rather small in health.[6] The makeup of the friendly microbes in the large intestine varies from one individual to another and is related to the environment and to the individual's genetic load, nutrition, and gender as well. Without these microbes, we would not survive. Considering the possible damage that could be caused to this population of microbes, we must be cautious in prescribing large doses of antibiotics. That said, some of the microbes in our gut can pose a real threat to health, as I'll discuss below.

We live in complete symbiosis with the microbes in our body in general, and particularly with those in our digestive tract, which are called gut microbiota. Other parts of our body are also host to unique microbiota, such as the skin, the nose, the sexual organs and more. Taken together, the different microbiota are known as the human microbiome. In this chapter we will not deal with the complete human microbiome but rather with the microbiota of the digestive tract. So where do the gut microbes come from? For over a century it has been thought that the fetal gut is sterile. In recent years mounting evidence indicates that this is not the case and that bacterial colonization of the gut occurs already partly in intrauterine life. Subsequently and based on the mode of delivery, breastfeeding and more, the neonatal gut microbioma is established during the first three days of life but it changes constantly across our life. The changes may be due to diet, stress, hormonal environment, infections, and medications.

The composition of the gut microbiota is unique to each human being, to the point where an individual's genetic fingerprint, it could be argued, should include his or her microbiota. Indeed, researchers at Tel-Aviv University have proposed an evolutionary theory they call the Hologenome Theory of Evolution. They argue that the human genome together with the genomes of the human microbiota should be seen as one evolutionary unit, as the connection between an animal or a plant and its microbiome is so profound that in effect they constitute a single evolutionary mega-organism. In this context, the microbioma plays the role of an essential "organ." This is also the basis why the researchers are suggesting a more holistic view of health and illness.[8] If we consider the number of cells in our mega-organism, we are 33% human and 66% microbes. Yet the gut microbioma contains approximately 5 million unique genes while human cells contain about 20,000–30,000 genes. Looking at humans and their microbes as forming one genetic unit, from the genetic perspective we are less than 1% human and 99% microbes.[9] Looked at that way, it's clear that any change in the makeup of the microbe population will have far-reaching implications.

Obviously, the food that we eat, the way we consume it, the amount we consume, and when we eat all have a major impact on not only our health but on our microbiome. Consider artificial sweeteners, one of the most widely used food additives. Most of these substances are indigestible by the small intestines and reach the large intestine and the microbiota virtually unchanged. In a recent study published in the journal *Nature*, researchers fed mice with artificial sweeteners in doses corresponding to the acceptable daily intake in humans. The

results indicated that artificial sweeteners might alter the microbiota resulting in an intolerance to glucose. Similar results were
also found in humans.[10] In other words, by using artificial
sweeteners we may increase our risk for diabetes.

Another example is fiber. In recent years, fiber supplements have become popular, as fiber is known to have the
ability to regulate the consistency and frequency of bowel
movements. Yet adding fiber to our diet has other, more important aspects. In general, the digestive system of mammals is not
designed to break down nutritional fiber, and certain types of
fiber such as lignin stay mostly intact but play a role in activating the intestine. They various types of fiber serve however
as an important source of nutrition for the microbiota in the
large intestine. Researchers in Switzerland have reported on
a mechanism that may explain how a diet rich in fiber affects
inflammatory processes in different regions of the body as well
as respiratory allergies.[11] Because such a diet changes not only
lung microbiota but also the composition of gut microbiota,
the concentration of certain fatty acids in the blood rises, which
consequently protects against inflammatory and allergic reactions. A high concentration of these fatty acids was found in
the blood of mice whose nutrition included dietary fiber, while
mice fed a diet low in fiber came down with allergic diseases
of the lungs.[12] It's increasingly clear to what extent diet influences our gut microbiota.

Yet from a gender medicine perspective, there are subtle
differences in how diet affects the microbiome. It has been
shown recently[13] that identical diets given to males and females
caused distinct changes in the gut microbioma for both sexes.
This held true for certain types of wild living fishes and for humans. The changes included the prevalence of different species
of microbes and changes in the diversity of certain microbial

sub-populations. Since these characteristics are to a large extent dependent on diet, the awareness that the same diets may have different effects on the microbiota in men and women may make it necessary to design special diets for men and women in order to treat the same disease.

Microbiota also affects the lifestyle of the host body. Take smoking for example. People who stop smoking often gain weight, and this weight gain is usually attributed to increased appetite. Yet it seems the reality is more complicated. In another study published in Switzerland, 80% of those who stopped smoking gained an average of seven kilograms without any change in the number of calories they consumed. The scientists, however, found changes in the composition of the microbiota of those who stopped smoking: after quitting their habit, the subjects' microbiota became enriched with species of microbes that are also found in high concentration in overweight people.[14]

What's more, the health of the microbiota is essential for the normal functioning of our immune system, which in turn safeguards the delicate balance between us and the various microbes, provides protection against invaders, and maintains the proper ratios between different microbe populations.[15] Disruptions may cause an imbalance in the immune system and lead to conditions in which the immune system attacks the host, in other words, to autoimmune diseases, inflammatory diseases, and allergies, all of which are more common in the Western world in general and among women in particular. These conditions are often related to Western eating habits, to overuse of antibiotics, and to exaggerated aspirations for cleanliness, which prevents our immune system from coping with the challenges of a decidedly non-sterile environment.

Studies conducted in the United States among the Amish,[16] a relatively closed agricultural population, and in agricultural regions in Switzerland and Austria revealed that children born into agricultural communities and exposed to a wide variety of plant and animal microbes—what modern society defines as filth—are more protected against allergic diseases such as asthma throughout their lifetimes than their cosmopolitan counterparts. Early exposure to environmental microbes is apparently necessary for training our immune system to distinguish between microbes that are harmful and those that are not. Exaggerated concerns for cleanliness apparently disrupt this adaptation—more reason to avoid unnecessary antibacterial soap.

The results of a study recently published in the journal *Science*[17] demonstrate the complex nature of the reciprocal relations among gender, the hormonal and immunological systems, and the microbiota. The research was conducted on a particular strain of diabetic mice. These mice, known as Non-Obese Diabetic or NOD mice, tend to develop autoimmune diseases spontaneously, which destroys the insulin-producing cells in their pancreas. Subsequently, their diabetes progresses to type 1 diabetes, a serious disease requiring ongoing treatment with insulin. The process by which the immune system destroys pancreatic cells is similar in these mice and in humans. The researchers observed that among these mice, exposure early in life to certain gut microbes led to an increase in testosterone levels and slowed down the progression to type 1 diabetes. Furthermore, transferring the microbiota from adult male mice to immature female mice provided them with similar protection. This study demonstrates two important mechanisms. First, the microbiota can influence our endocrine system by causing an increase in testosterone. Second, this study provides additional evidence that testosterone has the ability to suppress the immune system.

Not all microbes are beneficial, however. Our large intestine is home to some dangerous strains, which can lead to severe disease including allergies, autoimmune disease, bowel disease, colon cancer and more. It is only the broad variety of microbes there that prevents them from taking control, providing a certain equilibrium. For this reason, taking antibiotics in an unsupervised manner carries the risk of damaging a delicate balance of microbe coexistence in the digestive tract.

In all likelihood, a disruption in the balance between the host's body and its microbiota has a major impact on our health, perhaps no less than the human genome itself. Personalized medicine needs to widen its focus to include the microbiome, especially as our ability to influence the microbiota is certainly greater than our ability to edit our genome.

For example, recently there has been increasing interest in research on microbiota transplantation—the transfer of fecal matter from the large intestine of a donor to that of a recipient. This method was first described over 1,700 years ago, when physicians treated patients with fecal transplant given by mouth for severe diarrhea or food poisoning. Today, the way of performing fecal transplants and our ability to monitor results has grown more sophisticated due to advances in laboratory methods.[18] In recent years the method has been tried in patients with chronic intestinal infections caused by the clostridium bacterium and has achieved significantly better results than the usual antibiotic treatment, particularly in cases when the infecting bacterium had developed resistance.[19] As far as medical treatments go, fecal transplants are a relatively simple procedure. The transplant is prepared in the laboratory and the resulting liquid is then introduced into the intestines with the help of colonoscopy or by inserting a tube through the anus or through the mouth. Although data related to this therapeutic option are still scarce, there is growing optimism related to future developments in the procedure.

Returning to the effect of gender on the microbiota: It has been demonstrated by scientists[20] from Michigan, USA, that the development of our microbioma is dependent on many variables, amongst them life-history characteristics, breastfeeding as infants, level of education and of course gender. Once developed uniquely for each individual, the microbioma is subject to constant changes as result of lifestyle, diet, medications, stress, hormonal changes and actually most situations which affect us, including even jet lag. In women, these changes are even more prominent due the constantly changing hormonal milieu in the fertile age. The fluctuations in the composition of the microbioma are evident along the menstrual cycle and even more during pregnancy. In mice, the difference between the microbiota of males and females begins to appear during the maturation period and is affected by the presence of testosterone. (Castration stops the development of this difference.[21]) In humans, it was found that the microbiota of pregnant women undergoes numerous changes, and during the third trimester these changes are likely to manifest in symptoms that resemble metabolic syndrome, including resistance to insulin and weight gain. During this period such changes are beneficial; they support the growth of the fetus and prepare the mother's body for the caloric demands of breastfeeding.[22] In animals that are not pregnant, such changes are an expression of disease. Interestingly, microbiota taken from pregnant women during the third trimester and transplanted in mice whose gut bacteria had previously been destroyed caused the mice to gain weight and become resistant to insulin. These studies robustly demonstrate the impact of the microbiota on the body's metabolism—and one reason for the microbial differences between men and women.

The impact of the microbiota, however, is not limited to health and illness but also has an effect on various behaviors, at least in mice. Another study showed that when intestinal microbes from fat mice were transplanted into thin mice, it caused the latter to eat more.

The microbiota may even have an impact on mating preferences, at least in flies. Researchers at Tel-Aviv University fed two groups of flies with different types of food. After only one generation, the two groups began to prefer to mate with flies from the same nutritional background, a behavior known as mating preference.[23] In order to discover whether the microbiota of the flies was related to this preference, in a second stage of the experiment the researchers administered antibiotics to the flies, destroying their microbiota. The preference disappeared. It is interesting that this mating preference was created within a single generation but was preserved for 37 generations. From this we can conclude that at least among flies, the microbiota affects the choice of a mating partner and that this preference is inheritable.

The importance of microbiota research cannot be exaggerated. Many studies that attempt to look to genetic changes to explain various health phenomena are likely to benefit from more in-depth investigation of the microbial world inside our bodies. It is not surprising that the American National Institutes of Health (NIH) initiated a large research project known as the Human Microbiome Project as a continuation of the Human Genome Project, at a cost of 140 million dollars.[24]

The human digestive system is an extremely complex mechanism, and not all intricacies of its functions have been elucidated. The gender differences in digestive function, in particular, require further research. But at least this is a start.

CHAPTER 8

GENDER ASPECTS OF THE REPRODUCTIVE SYSTEM

A<small>S IS TRUE OF ALL LIVING BEINGS, THE EVOLUTIONARY GOAL</small> of the human species is to reproduce and to perpetuate its genes from one generation to another. The physiological differences between men and women have, naturally, given rise to the development of different reproductive strategies. A man produces around one hundred million sperm cells per day from the outset of his reproductive maturity and throughout his life, with some decline in quality and quantity. With each ejaculation a man releases between 100 million and 300 million sperm cells. Such a tremendous number could theoretically fertilize every fertile woman in Europe with a single ejaculation! This mechanism of constant sperm production contributes to the male's evolutionary reproductive strategy: to disperse his sperm and his genes as widely as possible without being particularly selective. Genghis Khan famously followed this strategy to its extreme: a 2007 report stated that based on genetic data, his descendants today number 16 million people.[1] Given this strategy—and the high chances of success—male responsibility for his descendants is negligible. Among most mammals, including human beings, responsibility for the descendants falls primarily on the female.

A woman's reproductive strategy is entirely different. First, a woman does not have a mechanism for producing eggs during her period of fertility. Rather, they develop while the female fetus is still in the womb—in utero she will produce around seven million eggs. Most of these atrophy before the female baby is born, and at birth only around a million and a half eggs remain in the ovaries. This degeneration continues until puberty and the onset of the menstruation, at which point the ovaries contain only around 500,000 eggs. With each ovulation, another one or two thousand eggs degenerate, until the supply is depleted, marking the end of the woman's fertility and the beginning of menopause. For this reason, the older the woman is when she becomes pregnant, the larger the risk of chromosomal disorders in her descendants—not because of the woman's age but because of the age of her eggs. The fertilized egg of a forty-year-old woman is forty years and several months old because it was produced even before she was born.

A woman's period of fertility is limited, the number of descendants she can produce is small, her pregnancies last for a relatively long period of time, and she is particularly vulnerable during and after her pregnancies. Due to her major investment in the reproductive process, a woman must be particularly selective in choosing a partner and in appraising his genetic quality, his ability to support and protect her during and after her pregnancy, and his willingness to safeguard their relationship and their descendants.

Such divergent agendas with respect to reproduction are of major importance to the social structure of humans and of animals. Of course this does not take into consideration contemporary societal changes among various population groups in the West which promote equality for women. Yet even a new social structure marked by a more equitable division of roles does not negate the basic biological difference that underlies reproductive relationships between men and women.

From the aspect of gender medicine, these different strategies of reproduction and the associated differences in anatomy and physiology require different diagnostic and therapeutic approaches when not functioning properly and lead to different expressions of disease in men and women. I will expand on these issues later in this and in the next chapter.

COURTSHIP

Numerous mechanisms are responsible for activating the reproductive instinct, among them sexual desire, tools for identifying the most compatible partner, effective means of courtship and tools for building and maintaining relationships. Among mammals, courtship is accompanied by an increased sense of energy and by focused and ongoing attention. It is difficult to ignore the similarity between courtship rituals among mammals and even among birds and expressions of love among humans, yet the differences between us are pronounced. An animal chooses a mate through the controlled evaluation of a particular individual according to a distinct set of criteria. When a human falls in love it feels fated, out of our control—we are instantly convinced that the object of our love is unique and extraordinary and irreplaceable. Yet, according to American anthropologist Helen Fisher, romantic love is an advanced technique of the mammalian brain for selecting a partner.[2]

It is generally thought that males are the suitors while females are sought after. Yet reality is much more complex. For male courtship to succeed, the woman must first convey, using body language, that she is willing to be courted. Some estimate that in 90% of cases it is the woman who consciously initiates the courtship in this manner. In their popular book on body language, Allan and Barbara Pease describe thirteen common body language messages that women use to transmit signs of willingness to be courted.[3] The man must then read and react to such messages appropriately before a proper courtship can begin.

The courtship process is accompanied by physical and neural events, which modern science has given us tools to identify and analyze. We can trace various chemical substances secreted by the brain during courtship and sophisticated imaging techniques allow us to identify the regions of the brain activated by these mechanisms. For example, among some mammals courtship actions are accompanied by a rise in the level of dopamine, a neurotransmitter that transmits signals between neurons in certain regions of the brain. This increase, which is related to enhanced attention, activates the classic symptoms of falling in love: it raises pulse and blood pressure, improves mood, and induces an overall sense of contentment. Interestingly, stress hormones such as cortisone are also implicated in the success or failure of courtship. Women are less receptive to courtship when under stress, as opposed to men with high levels of cortisone who become more receptive.

When courtship succeeds, a couple will get to know each other via verbal and nonverbal means, behavioral means, and the operation of all five senses. The biochemical basis of some of these means has been thoroughly studied. Take for example the role of sexual kissing in the process of selecting a partner. Apologies to the romantics out there, but kissing is not only romantic. It conveys information for mutual assessment of mating quality.[4] During sexual kissing, the brain releases dopamine and other hormones, such as oxytocin, a hormone involved in giving birth and breastfeeding. These two hormones trigger a feeling of attraction and pleasure.

In addition, the composition of the saliva is different in

men and women and the exchange of fluids between the two kissers transmits important information that helps them evaluate the unique qualities of their partner. For example, in sharing her estrogen, the woman transmits a message to the man regarding her fertility. The composition of a woman's saliva also changes throughout her menstrual cycle under the influence of estrogen. Around ovulation, her saliva contains more sugar and her kiss is sweeter in all senses of the word. This may be an evolutionary tool to encourage mating at the time when chances for impregnation are highest. The testosterone which a man passes to a woman through his saliva is meant to enhance her willingness to have sexual intercourse and to signal his sexual prowess. In addition, it is likely that a woman can "read" her partner's saliva in order to determine his immunological status and his resistance to infectious diseases. The physical proximity while kissing also facilitates more effective use of the sense of smell and the interpretation of messages regarding a partner's general health. A study conducted in the United States indicates that more than half the women who were attracted to a particular partner discontinued the contact between them after a few kisses, likely because she unknowingly assessed him to be incompatible with her, healthwise, and not necessarily due to deficient technique.[5] Kissing was shown to serve men more as a means of advancing their chances of having sex and women as a means of helping them choose a suitable partner. Most of the men in the study, which included more than a thousand participants, stated that they would have sexual intercourse with a woman they had not kissed previously, but only one out of seven women indicated they would do so with a man they had not previously kissed.

Sexual desire and romantic love are related to each other but are certainly not the same. Indeed, they even look different in the brain: the brain centers responsible for sexual desire and those related to expressions of love are located in separate regions, and are even positioned differently in men and women. In a telling study, researchers from the USA and Switzerland used sophisticated imaging techniques to examine groups of men and women in love, showing them, among other things, photographs of the object of their love.[6] Increased brain activity was observed among men in the regions responsible for processing visual messages, while among women increased activity was observed in brain regions responsible for memory, attention, and feelings. Whoever so desires can see in this an explanation for the fact that men are more attracted by visual qualities like youth and beauty, while expressions of attention and sensitivity hold more importance for women. This means that men are more attracted to how the object of their sexual desire looks and women more to how it acts. These differences impact on and serve the different reproductive strategies, which I discussed above. Human sexology is a further extension of our topic and deals with sexual function and dysfunction and with the important gender differences in this area. We will not discuss this topic here and we will restrict ourselves to the reproductive process—the most amazing process in nature.

THE REPRODUCTIVE PROCESS

It's incredible that a simple merger of two cells—sperm and egg—can, within a period of nine months, result in a new human body containing ten trillion cells. Moreover, all the cells in our body carry the same, original genetic load, but combine to produce numerous complex systems. This miracle is the result of the reproductive process.

In fertile women, thousands of follicles grow in the ovaries each month. One (or sometimes more) of these follicles bursts and releases an egg into the fallopian tube which leads to the uterus in a process called ovulation (see Fig. 1). The egg waits in this tube, ready to be fertilized. If the egg is not fertilized within a short period of time—12 to 24 hours—the window of opportunity closes and the egg is no longer viable. Because most couples do not have daily sex, it would require a great coincidence for sperm to be present in the fallopian tubes at the required time. Hence, a compensatory mechanism emerged in nature. Sperm cells evolved greater survival abilities and can remain in the female genital tract for over a week after intercourse awaiting an egg. Thus a woman can in effect maintain a reservoir of sperm in her cervix for up to a week. This mechanism also explains why there are no "safe days" for preventing pregnancy prior to ovulation.

The tadpole-like sperm cell is unlike any other cell in the body. It consists of a head containing genetic material, a neck comprising mainly energy-producing organelles, and a tail that translates this energy into the movements typical of sperm cells. The mission of sperm cells is to exit the male body into that of the female, to travel against the current of mucus flowing out of the cervix, to cross countless barriers and to arrive at the coveted goal, the single egg in the fallopian tube. Of the hundreds of millions of sperm cells embarking on this journey in a single ejaculation, in the best of cases only one sperm cell will manage to fertilize the egg—the victory of one out of hundreds of millions. From this perspective, each one of us represents a winner.

Eventually, a group of sperm cells will encounter the egg that has just been released from the ovary. The egg proceeds slowly toward the uterus, moved by wavelike contractions of the fallopian tube, wrapped in designated cells known as granulosa cells. The encounter between the egg, the body's largest cell, and the sperm cells, the smallest, is dramatic. Like enthusiastic fans attempting to push toward the entrance of a stadium, the sperm cells proceed tirelessly in an attempt to penetrate the cell layers surrounding the egg. But this is not merely a mechanical process of active sperm cells and a passive egg. It is a complex cooperative process involving catalyst substances called enzymes. This cooperative action leads to the creation of a channel through which the sperm cell travels to the egg, and after one sperm cell succeeds, it blocks other sperm cells from continuing to dig through the surrounding cellular envelope. The winning sperm cell then "closes the gate" behind it. The merger between sperm and egg has begun, together with the creation of a new life.

Now let's return to anatomy.

THE OVARIES AND THE TESTICLES

The testicles are located in the scrotum, a sac of skin outside the abdominal cavity, an ostensibly vulnerable spot. They are located there because the sperm production process requires a temperature approximately two degrees lower than our core body temperature. The ovaries, in contrast, are found within the abdominal cavity, in a seemingly protected region. Yet the actual picture is the opposite. The testicles, despite working tirelessly like a conveyor belt producing astronomical numbers of sperm, stay intact through this accelerated activity. Ovulation, in contrast, is a process harmful to the capsule of the ovaries. An egg-containing follicle matures within the ovary, and releasing the egg necessitates a rupture of the ovarian capsule. Some women experience pain upon ovulation, others, bleeding. Each ovulation leaves a scar on the ovarian wall. This recurring scarring and healing can have grave consequences. In fact there is a direct relationship between the number of times a woman ovulates during her lifetime and her risk of developing ovarian cancer. Pregnancies and extended periods of lactation during which a woman does not ovulate significantly lower her risk of ovarian cancer, as does the use of birth control pills, which suppress ovulation.

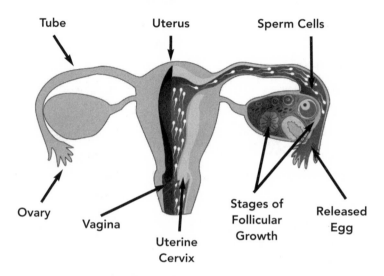

Figure 1: Anatomy of the female reproductive system during ovulation.

The location of the ovaries also has implications for women's health. The testicles are located outside the abdominal cavity, as are the tubes of the sperm release pathway. In men, the abdominal cavity is not connected to the outside world and under ordinary circumstances infectious agents such as bacteria have no access to it. This is not the case for women. Sperm cells arrive from the outside world through the vagina, the uterus, and the fallopian tubes, which open into the abdominal cavity. The passage intended for the sperm cells is often exploited by various bacteria and microorganisms. This can cause a disease that is virtually unknown among men—Pelvic Inflammatory Disease (PID), which can be quite serious and may lead to severe pain, fever, the need for repeated hospitalizations and even to an increase in ectopic pregnancies or infertility.

Women's health is also the central concern when talking about birth control. Since ancient times women have been thought to be responsible for reproduction, for birth control, and even for the sex of the newborn. Even today, in some countries a man is entitled to divorce his wife if the couple does not manage to have children. This is also the case if the woman does not produce a male child, though there is no doubt today that it is a chromosome from the male that determines the sex of the fetus. Likewise, the consequences of an unwanted pregnancy must be borne by the woman.

The invention of the birth control pill in the mid-twentieth century gave women an effective means of family planning and constituted an important turning point in the battle for women's liberation. Yet even today, with the exception of the condom, the majority of birth control methods are intended for use by women, among them the birth control pill, the Intra Uterine Device (IUD), and tubal ligation. Vasectomies, minor operations aimed to disrupt the sperm conveying tubes, are common in very few countries. The reasons for this situation are numerous and include scientific and social considerations. Psychological gender factors may also interfere when it comes to the acceptability of birth control for males. What do I mean by this? With the exception of castration through surgical removal of the testicles or ovaries or sterilization (disruption of sperm-conveying structures in the male or ligation of the tubes which connect between ovaries and uterus in the female), there is no birth control method that is one hundred percent certain. If a woman has used birth control and becomes pregnant it will undoubtedly be accepted that the pill (or IUD, etc.) failed. But if a man uses birth control and his partner gets pregnant, the man may suspect that he is not the father of the child with all the resulting consequences on the relationship of the couple.

• • •

Human reproduction is a fascinating and central topic in medicine. Contraception and much more so, treatment of infertility, aim at a common goal in which two individuals are involved. Doctors are therefore confronted with a unique situation in which they simultaneously assess and treat two patients who differ by sex and gender towards a common desired outcome. Therefore, many aspects of gender medicine affecting different bodily functions may materialize in either of the partners. In the next chapter I will discuss gender aspects of infertility without entering too deep into the technical detail of treatment of infertility.

CHAPTER 9

GENDER ASPECTS OF INFERTILITY

A S I ILLUSTRATED TO SOME EXTENT IN THE LAST CHAPTER, the human reproductive process is complex. It's becoming increasingly clear that the differences between the sexes are not only physiological. They involve value systems, cultural, anthropological, and psychological issues, sickness and health, and numerous other areas as well, and cause men and women to react differently to any given diagnosis or treatment. Nowhere is this more evident than in the case of infertility.

The World Health Organization (WHO) defines infertility as a disease of the reproductive system marked by the failure to achieve a clinical pregnancy—the presence of an embryo sac inside or outside the uterus—after 12 months or more of regular, unprotected sexual intercourse.[1] As with any disease, implicit in this definition is the basic right to treatment. Yet unlike other diseases in which treatment is directed at an individual, in the case of infertility, treatment is necessarily directed to two individuals—the woman providing the egg, and the man providing the sperm. This is true even for women undergoing artificial insemination from donor sperm. Though this adds a level of complexity in the diagnosis, the dual diagnosis has some upside: if fertility is a combination of the reproductive function of both members of a couple, in many cases the proper functioning of the reproductive system of one partner can compensate for the less than optimal functioning of the other.

Around 5% to 10% of couples in the West meet the definition of infertility. (I am intentionally avoiding the common term sterility, with its implied irreversibility and associated stigma.) Infertile couples often find themselves under tremendous family and social pressure that in some societies reaches intolerable proportions. In many cultures, blame is directed toward the woman. Moreover, infertile couples endure self-directed pressure as well. Hopeful parents' reactions to the bitter news of infertility range from shock to depression, emotional stress, frustration, and a decrease in self-esteem. These responses find expression in different forms and degrees in each member of the couple.

In general, the reasons for infertility are associated with the woman in 30% of the cases, with the man in 30% of the cases, with shared factors in 20% of the cases, and with no apparent reason in 20%. Yet each case is unique, and awareness of such percentages cannot soothe a partner who finds himself or herself the source of the problem. In such cases couples find themselves having to cope with feelings of guilt and mutual accusation, and their tremendous emotional burden is not difficult to imagine. To make matters worse, the level of invasiveness varies between treatments for the two sexes. In men, the basis test for fertility is typically an examination of sperm samples. For women, the testing procedure is more complex and includes ovulation testing, testing for mechanical factors to determine whether the egg can pass freely into the uterus and whether the sperm can reach the egg, and testing for additional and rarer factors.

This imbalance is just one of many gender-based factors that can complicate the treatment process for infertility. But by more closely examining the mechanics of the causes of infertility and its implications, I hope to further the conversation of how to more effectively and sensitively treat couples going through this inherently difficult diagnosis.

OVULATION AND SPERM CELL PRODUCTION

Because the mechanisms of ovulation and of sperm cell production are so different, the therapeutic approaches for treating each differ. In this chapter I will not discuss specific methods for treating infertility, because of their more technical nature. I will, however, briefly discuss the differences between ovulation and sperm production, the therapeutic implications of these differences and the emotional aspects of infertility treatment.

Ovulation is a cyclical process that begins with puberty and ends with menopause. The ovaries have a limited reservoir of follicles containing eggs produced while the woman was still a fetus in her mother's womb. During each cycle, a group of follicles mature and the eggs mature within them. Ultimately one follicle (or sometimes more) bursts, releasing the egg into the Fallopian tube, a delicate structure leading to the uterus. The encounter between the egg and the sperm cells takes place in the Fallopian tube, and the result of this encounter is fertilization by a single sperm cell. After several days, the resulting embryo reaches the uterus, implants there, and the pregnancy begins.

Given the complexity of the process, any number of things can go wrong. Due to an anatomical defect in either partner, the seminal fluid containing the sperm cells may not reach the vagina or the cervix. The sperm cells that reach the vagina may be inferior in quantity or in quality. The passage from the vagina to the uterus and from there to the Fallopian tubes may be damaged or blocked. And even if the sperm cells reach their destination, they may not find an egg there due to impaired ovulation. Given these odds, it's a miracle that any eggs are ever fertilized at all.

The cyclical nature of ovulation is thanks to the endocrine system. This system is made up of a number of glands which function as if on an axis, with each activating the next station, controlled by regular feedback (Figure 1).

Figure 1: The endocrine cycle that controls the ovaries and the testicles

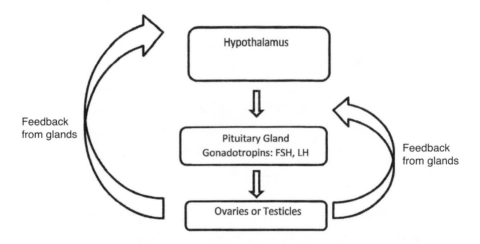

The highest station in this glandular cycle is situated in a region of the brain called the hypothalamus. The hypothalamus secretes a gonadotropin-releasing hormone, which in turn activates the pituitary gland, a small organ located in the center of the brain. Among other things, this gland secretes two hormones, the gonadotropins LH and FSH, which activate the ovaries in women and the testicles in men. In women, FSH causes the follicles to mature and at the same time causes the secretion of the hormone estrogen, while LH causes the follicle to burst—an event known as ovulation—and causes the secretion of another hormone known as progesterone. Each gland returns feedback along the axis and thus controls the secretion process.

All fertility treatments geared at causing ovulation are based on imitating this natural process or part of it. Hormonal preparations are administered in cycles with the goal of causing the follicle and its egg to mature. For In Vitro Fertilization (IVF), the therapeutic target is somewhat different. Here the objective is to cause a large number of follicles to mature at once while observing their maturation by means of ultrasound imaging, to aspirate the eggs using a special needle while the woman is under some form of anesthesia, and then to fertilize them under laboratory conditions. If successful, one or two embryos are transferred to the woman's uterus using a thin catheter, and the remaining embryos, if any, are preserved by deep freezing to be used in future cycles. To achieve accelerated maturation of a number of follicles, intensive hormonal treatment is required, with numerous side effects. Hormonal treatment protocols for inducing ovulation are rather effective whether or not combined with IVF. It is estimated that since the birth of Louise Brown, the first "test tube baby" in July 1978, close to 6 million children have been born worldwide as a result of artificial reproductive technologies.

Sperm production on the other hand, is ongoing. As noted, sperm cells are produced regularly and in tremendous quantities (see Chapter 9). The process takes place inside convoluted tubules located in the testicles and is controlled by a similar hormonal system as that in women. The hormone FSH prompts production of the sperm cells, and the hormone LH causes testosterone to be secreted by special cells scattered between the sperm cell–producing tubules (Leydig cells). Yet in men these hormones are secreted at a more or less constant— not cyclically. This secretion pattern originates in the hypothal- amus. By default the hormone that activates the pituitary gland to secrete gonadotropins is secreted cyclically. In men, begin- ning in utero, high levels of testosterone (see Chapter 3) perma- nently suppress the cyclic secretion pattern leading to uniform secretion. Animal experiments have shown that blocking this process leads to the birth of males that do not produce sperm. In contrast, blocking the cyclical center in females by exposing them to high levels of testosterone leads to the subsequent sup- pression of ovulation. For this reason, hormonal treatment in men aiming to produce sperm is administered continuously. Unfortunately, and despite attempts to imitate nature as much as possible, these treatments have not been very effective. Hence, men with infertility problems are currently almost never treated with drugs.

Although the procedure for determining the causes of male infertility is relatively simple, treatment is not very reward- ing. Diagnosis is usually based on examination of the semen, including sperm count and the motility and structure of the sperm. Yet with the exception of a few pathological conditions that can cause a decrease in these measurements, in most cases no explanation is found. Even if they are, there are no effective medications for increasing sperm count or improving sperm quality. Therefore, the most common method of treating male fertility is currently to bypass rather than to correct the prob- lem. In the bypass method, whatever can be retrieved from the sperm sample or from the testicles is brought into as close con- tact as possible with the egg, whether by transferring sperm cells directly into the uterus (intrauterine insemination, or IUI) or by means of in vitro fertilization. Modern techniques for inserting a single sperm cell directly into the egg have proved to be very successful. Surgical techniques have also been developed for retrieving isolated sperm cells by puncturing the testicles and using these to fertilize the egg.

Ultimately, no matter the method, the physical and emotional burden of fertility treatment, including hormonal preparation, is borne by the woman. In other words, even male fertility treatments are for the most part implemented in the woman's body. Not only that biologically women are the ones who must bear and develop the fetus for nine months, they are also the ones who must undergo demanding hormonal therapy. Artificial insemination takes place in their uterus. And in the case of innovative new technologies, including in vitro fertilization, it is the woman who must undergo complicated hormonal treatments, frequent hormone tests, egg retrieval requiring anesthesia, and transfer of the embryos into her uterus. All of these treatments have side effects, some severe, and this is in addition to the emotional stress that in itself can reduce chances for success.[2]

THE EMOTIONAL BURDEN OF
FERTILITY TREATMENTS

As a rule, women are under more emotional stress in preparation for in vitro fertilization than are men.[3] Yet, the male partner must also carry his part of the emotional load. Some will say that little is being asked of him, that all he has to do is submit a few sperm samples. But we should not belittle this.

Just for the sake of illustration, usually the sperm sample must reach the lab in the morning and therefore be collected early. The alarm that wakes the man up at six in the morning to "produce" sperm turns sex for both partners into a mechanical process, and this is not at all a trivial matter. Men undergoing testing to diagnose infertility exhibit an increased rate of sexual dysfunction, such as early ejaculation and problems achieving an erection. In a study in the United States of 121 men with infertility problems, 23% reported depression and 22% reported erection problems,[4] which is comparable to women undergoing testing for infertility work, who also exhibit a high incidence of depression and sexual dysfunction. Various degrees of depression were diagnosed in one-third of 121 women who took part in a survey, and one-fourth reported negative influences on their sex life.[5] In developing countries, women suffer disproportionately from the stigma of infertility, and this has severe social ramifications. However, in developed nations, stigma surrounding male infertility might be even more severe than those directed at women. Many men do not sufficiently understand the biological basis of infertility and are likely to confuse infertility and impotence. To observe men sitting in the waiting rooms of male fertility clinics and to note their embarrassment, their desperate attempts to avoid eye contact with others, or to listen to them when describing their symptoms is to understand the depth of their emotional vulnerability. Add to this the inability of many men to express their feelings in words and share their emotional problems and we can better understand their emotional distress. Women and men experience infertility differently on an individual basis and as a couple, they employ coping strategies differently and react differently to the coping mechanism of their respective partner. In a large study researchers from Denmark examined coping strategies in over 1,000 infertile couples[6]. They described that the basic coping strategies employed are active avoidance (avoiding children, pregnant women or discussions related to these), passive-avoidance (hoping for something miraculous to happen), meaning based strategy (re-organizing life, in spite of infertility and finding new goals) and active confronting (expressing feelings, talking with others about the problem). Women must undergo difficult treatments, endure side effects, generally suffer a higher negative impact on their quality of life and experience infertility as a greater threat than men.[7] It is probably therefore that women use all of these coping strategies more intensively than men and most often they use active confronting.[8] Men on the other hand use more distancing, try to plan actions and invest in self-control. Obviously, the coping mechanism of one partner has an effect on the infertility stress on the other partner. Active avoidance in one partner increases infertility distress in the other. If men use more distancing, while their partner does not, their stress level increases. Active confronting coping by women increases infertility distress in the male partner, that is, when women choose to share their stressful experience with others, this actually increases the stress in their male partners and the marital stress. The use of meaning-based coping also has different impacts for men and women. When men turn to this strategy, the social stress in their partner actually increases but when women use this strategy, men experience a decrease in marital stress.

Because the situation is a significant shared stressor for both partners, the coping strategy of either partner will affect the other and may have a considerable impact on treatment results. Emotional calm and mutual support are essential to the success of any infertility treatment. A case in point is that of artificial insemination using donor sperm. When the male partner is diagnosed as infertile based on a total lack of sperm cells in the semen, sperm cells can sometimes be retrieved by puncturing the testicle. If this does not work, a last option is artificial insemination with donor sperm. A major research study published in Israel 30 years ago[9] examined how mutual support by the members of the couple affected the results of artificial donor insemination treatment. Two hundred and seventy couples treated by the same physician under the same conditions went through a preparatory process which included explanation of the nature of the treatment and its stages and signing informed consent forms certifying that each partner understood and agreed to the process. The men were asked to accompany their female partners to the insemination sessions, to be present during the insemination, and even to play an active role, including operating the syringe transferring donor sperm to the vagina. The men were also offered the possibility of mixing their own sperm with that of the donor in order to introduce a "positive doubt" regarding the possibility that it was the husband's sperm that ultimately fertilized the egg. In 235 of the couples, the man fully cooperated with the process, while in 35 couples the man never showed up at the treatment sessions. The treatment results were markedly different between the two groups. In the cooperative group, the cumulative pregnancy rate was 93.6% and the miscarriage rate was 15.9%. In the other group, defined as the unsupportive treatment group, the pregnancy rate was only 28.6% and 30% of these pregnancies ended in spontaneous abortion.

The power of psychosomatic effects on fertility treatments is best illustrated by one case that forever remains in my memory. A young and well-educated couple had been infertile for five years due to an extremely low sperm count. Having become discouraged with various treatment methods, they enrolled in an artificial insemination program with donor sperm. Even though extensive testing of the woman revealed no problems, after eight or nine cycles of artificial insemination with donor sperm she still had not become pregnant, and the couple's frustration (and mine, as their treating physician) mounted from month to month. One day, after a year and a half of treatment cycles with breaks in between, the woman announced to me, before yet another cycle of artificial donor insemination, that she was sure that "this month it's going to work!" She told me, in words that I still remember almost 35 years later: "When we first came to you it was clear who was to blame for our infertility problem. If I had become pregnant in the first or second month, the blame on my husband would have always been hanging over our heads. Now, after so much time and suffering, it has become clear that I am also at fault. I am also responsible for our many failures. Today my husband and I equally share our problem. Therefore now we can succeed." That same month she became pregnant.

Even after the pregnancy is achieved, men and women have divergent emotional responses. Women tend to worry about the continued normal course of the pregnancy, while men are more concerned about the health of the unborn child. As in any area of medical treatment emotions must be taken into consideration, and this is even more important when it comes to fertility. For this reason, the world's leading centers for fertility treatment employ professionals in the field of psychology on their treatment teams out of an awareness that various expressions of emotional stress on the part of the couple and knowledge of coping mechanisms are important to the success of the treatment. This understanding is becoming more widespread, but there is still room for improvement. Clearly, women and men need to be treated differently for infertility because of biological differences. This applies also for the psychological aspects of coping with infertility. The main physical and psychological burden is on the female partner but both partners are confronted with a uniquely stressful situation, with which they cope differently.

Gender medicine addresses this issue by pointing to the need for gender-specific approaches. Infertile men and women, including same-sex couples, have special psychological needs during and between treatment cycles. Although many major centers for reproductive medicine have psychologists on their staff, their involvement is often restricted to initial screening and to complicated cases. It seems logical that psychological care which takes into account the different coping mechanisms of men and women should be routinely extended to accompany and provide continuous support for infertile couples. This service does not necessarily need to be given in the framework of the actual infertility treatment facility. As orthopedic surgeons refer their patients to physiotherapy or cardiologists to rehabilitation centers where specific care is given concomitant with medical care which they receive elsewhere, so infertile patients could be referred to specialized centers, where psychological support is offered on a continuing basis and parallel to their infertility treatment. I am not aware that such a model is currently active somewhere but it seems to me certainly worth of consideration.

This chapter cannot be concluded without pointing to major bias in reproductive medicine: Treatment of female infertility has made tremendous progress in the past few decades and it can be safely stated that techniques of ovulation induction, in vitro fertilization and other methods of assisted reproductive technologies have helped to overcome the vast majority of infertility problems in women. Unfortunately, treatment of male infertility has greatly lagged behind. Apart from surgical techniques to retrieve sperm cells directly from the testes, a technique applicable for an extremely small proportion of infertile men, no real progress has been made for many decades. Even this technique requires subsequent assisted reproductive technologies in the female. So, both for infertility treatment and for contraception the onus remains almost exclusively on women. This is a troubling situation which requires urgent reappraisal, new strategies in research and funding, and Gender Medicine aims at raising awareness on this topic.

CHAPTER 10

GENDER ASPECTS OF PAIN

M EDICINE DEALS WITH THREE MAJOR AREAS: SAVING LIVES, prolonging lives, and improving quality of life. Within the latter category, the treatment of pain is a central issue. One of the primary purposes of general medicine is to reduce and prevent pain. And while no one *wants* to experience pain, acute pain can be an important protective mechanism whose main physiological purpose is to warn us of imminent danger. If we place our hand on a hot surface, the sudden pain we feel will cause us to pull it back. Intensifying pain in the chest may signal a heart attack. Sudden pain in the eye can point to the presence of a foreign body. Pain in an injured limb can prompt us to limit its use and thus aid in the healing process. In this way, pain enables us to take immediate measures to avoid harm to our bodies. These essential warning signals also have major survival value. There exist rare conditions in which individuals cannot feel physical pain. These unfortunate people are susceptible to harm and injury as they don't receive the body's natural warning signals. They are repeatedly being hurt and their life expectancy is significantly lower than that of healthy individuals.

Not every sensation of pain is essential to our survival, however. In contrast to the positive role played by acute pain, chronic pain is non-adaptive and its purpose is unclear. Headaches, migraines, backaches, the pains accompanying cancer and other forms of chronic pain are all superfluous from the physiological perspective. Chronic pain sets off a vicious cycle: it leads to anxiety and depression, to unnatural body postures and awkward positioning of the limbs, to sleep disturbances, and all of these in turn amplify the sensation of pain. Indirectly, chronic pain can also aggravate illness. For someone who has had abdominal surgery, for example, breathing deeply is painful, and he or she may begin to breathe more shallowly, which can lead to other health complications. Such cycles must be broken and major research and therapeutic efforts have been invested in finding ways to do so.

Another pain category, positioned somewhere between essential and superfluous pain, is what I would describe as "rewarding pain." In such situations pain serves a desired end, such as muscle soreness during and after athletic activities. Probably the best example is childbirth and its associated uterine contractions, also known as labor pains. The women giving birth and their care givers are aware of the positive purpose of this pain and acknowledge that each painful contraction is inching closer to the positive result of the baby's delivery. Indeed, some women giving birth view this pain as a positive—albeit sometimes excruciating—experience. Of course, the same effect of labor could be achieved without the accompanying sensations of pain, and modern delivery wards are geared towards helping alleviate the pain associated with childbirth. Still, many women consciously choose to forego an epidural anesthesia or pain alleviating medications.

In medical terms, pain can be classified as severe or as chronic based on specific criteria: location (the lower back or the stomach); the type of tissue it affects (muscle or bone); how it spreads (focused or radiating); and the cause (after surgery or pain for which there is no apparent cause). And as I'll explain later, gender influences an individual's experience of each of these categories. However, neither of these facts changes the fundamental problem of pain in medicine: doctors must treat patients by trying to objectively diagnose what is ultimately a purely subjective sensation.

Assessment of pain is considered today as the fifth vital sign, in addition to pulse, blood pressure, body temperature, and respiration rate. Nonetheless, despite our insights regarding the nature of pain, clinical treatment of pain is often deficient. In 2005, a survey of more than 4,500 patients at a hospital in Italy showed that only one-third of those who complained of pain received painkillers.[1] This example and others point to the fact that many health care practitioners still cling to the outmoded idea that post-operative patients should be given painkillers only if they specifically request them. In contrast, the modern approach is to offer painkillers in advance and as a preventive treatment. Of course, preventive treatment for pain must be administered wisely so as not to mask the warning signs of pain that can appear following complications of any sort.

Pain treatment is complicated as it is, but as made evident from the childbirth example, women and men may be biologically subject to different kinds of pain. Which leads to yet another question: do the two sexes express and experience pain differently? Do they react different to treatment of pain? What can gender medicine teach us about how to appropriately treat pain? In this chapter, I'll discuss types of pain, how information about pain is transmitted to the brain, how painkillers work and the gender differences marking each of these. I will try to make a case for the need to view and treat pain in women and men differently and will show that not to do so puts women foremostly into an unacceptable disadvantage.

WHAT IS PAIN?

The International Association for the Study of Pain defines pain as "an unpleasant sensory and emotional experience associated with actual or potential tissue damage, or described in terms of such damage." Inherent in this definition is the subjective nature of pain and the acknowledgment that we lack the tools to objectively assess it. In order to quantify pain we must rely on subjective verbal reports, such as the patient's descriptions, as well as an observation of body language and facial expressions. Other indirect means are used to assess pain, such as patients' pain diaries, including a measure for rating pain on a scale from 1 to 10. Just as it is difficult to rate feelings such as anger, love, or longing, among other reasons because of the great variation between individuals, so too it is difficult to derive precise data on pain by means of subjective ratings. Moreover, it is difficult to describe pain in words.

It's all the more challenging to quantify pain when the subject has limited verbal skills, for instance children, cognitively challenged persons, or non-native speakers of the doctor's language; or lacks the ability to speak, for instance babies or animals. Even physiological measures such as pupil responses, elevated pulse and blood pressure, and measures of stress hormones are imprecise, again because of the great variances among human beings. In clinical situations, we rely heavily on facial expressions of pain. (As I'll discuss in chapter 15 the deciphering of facial expressions is crucial for social life and many brain resources are devoted to this activity.)

Yet here, too, we run into complexities of variance. Facial expression of pain and additional messages conveyed by body language are partly under voluntary control. Patients may have various agendas related to expressing pain—procuring the

170

maximum dose of painkillers, for instance. Health care personnel may have differing levels of sensitivity in decoding messages. Cultural and gender-related stereotypes make the task of decoding facial expressions of pain even more difficult. In order to talk productively about pain, then, we must first distinguish between sensations of pain, a biological phenomenon, and expressions of pain, which for the most part are the result of social factors including gendered, ethnic, educational, and behavioral norms.

The biological aspects of pain are related, among other things to age, the patient's health, endorphin levels (painkillers secreted by the brain), the patient's hormonal condition and genetic structure, and yes, the sex of the individual. For example, the influence of hormones can be seen in cases of migraines. When the levels of the female hormone estrogen rise, for example during pregnancy, there is often an accompanying drop in the severity and frequency of migraine headaches. In contrast, when estrogen levels drop, such as during menopause, we generally see an increase in joint pains. Genetic influences on pain can be illustrated by the experience of people with red hair. For genetic reasons, redheads are more sensitive to pain and are less responsive to anesthesia than brunettes or blondes, and this is more pronounced among redheaded women than among men. For this reason, they require greater quantities of anesthetics for any procedure. According to a study conducted in the United States, redheads avoid going to the dentist twice as much as other people for fear of pain.[2] Moreover, these redheads are more sensitive to painkillers that attach to KAPPA receptors, as discussed below.

In contrast to these biological aspects, expressions of pain are affected by your mental state including anxiety and depression, and by previous memories of similar pain. Hence, in an attempt to provide as accurate an assessment of pain as possible, we integrate various objective and subjective factors on which expressions of pain are based, with the result that we are less able to quantify the feeling of pain or the pain itself. In addition, the study of pain in living beings is limited (or should be limited) by ethical considerations. In recent years, automated systems are being developed which may help to more objectively assess facial expressions of pain.[3] Yet despite these drawbacks, today we have methods that are bringing us closer to a more objective assessment of pain. The functional MRI (fMRI) technique enables us to observe different regions in the brain that are activated during pain trials, for example in response to local heat.[4] Experiments using fMRI indicate that women, more than men, interpret heat stimulation as pain and that the brain regions associated with feelings of pain are activated sooner and more intensely in women. Additionally it has been found that brain regions activated during pain are located in different places in men and in women.

Like all physical sensations, the feeling of pain starts in the brain. The area of the body subject to pain—for instance, a hand reaching for a hot stove—transmits this stimulus to the brain. Particular brain regions locate and interpret this received information, compare it to a database containing previous incidents and construct what we feel as pain. Without this transmission of information, there would be no sensation of pain. Pain-relieving medications either act to block the transmission of information from the site of the event and along the path of information to the brain or obstruct the brain centers that interpret the information. Two main types of nerves are activated in transmitting this information, resulting in the diversity of pain we feel. One type, known as A-delta fibers, transmits information rapidly and produces a sharp and quick sensation of pain. The second type, known as C fibers, transmits the information more slowly and produces a sensation of burning and ongoing pain.

The substances that activate and connect neural fibers and open the flow of information to the brain are known as neurotransmitters. They report from the field to the neural fibers by attaching themselves, like a key to a lock, to receptors located at the edge of the nerve. Prostaglandin, one of the better-known neurotransmitters, also causes local blood vessels to expand and increases the flow of blood to the area. This is why we feel a sensation of heat, redness, and swelling in painful areas. In addition, prostaglandins can change the sensitivity of the C fibers that transmit the information about pain more slowly. In other words, the prostaglandins themselves can increase the sensation of pain and many pain-relieving medications are geared to suppress these neurotransmitters.

The third important type of long neural fiber is known as A-beta fibers. These fibers transmit information to the brain and to the spinal cord, but this information concerns not pain but other stimuli of the limbs and the body, such as touch, pressure, and motion. Interestingly, when we are feeling pain, different neural fibers apparently "talk" to each other. This is the reason that a painful incident in one area of the body may manifest as pain in a seemingly unrelated spot. For example, pain deriving from a heart attack may cause sensations of pain in the shoulder. However, the flow to the brain is limited by its bandwidth—it is capable of transmitting only a limited amount of information at a given time. The A-Delta and C pain fibers are in a sense competing with the A-beta sensation fibers. Put another way, if we transmit non-harmful information via the A-beta fibers, we have a chance of reducing the pain information transmitted via the C fibers. This is the reason why it is possible to reduce the pain or in essence disrupt the information being transmitted about pain in a finger that we have just hit with a hammer by pressing on or rubbing the finger. Here, then, is a scientific explanation for why a parent's loving caress on a painful spot helps diminish the pain, beyond the emotional effect of empathy.

Even when it comes to measuring pain threshold, that is, the degree to which human beings can tolerate pain, it is difficult, if not impossible, to separate and quantify the voluminous data. The question of pain tolerance becomes even more complex in view of increasing evidence regarding gender differences in degree of sensitivity to pain, some of which are dependent upon the hormonal systems in men and women. Sex hormones such as estrogen and progesterone influence pain and our response to pain both qualitatively and quantitatively.[5] It is well known that men's increased tolerance to pain is related to high testosterone levels. This makes evolutionary sense from our ancestral history: a male who spent the day hunting and fighting would be at a distinct disadvantage if he were bothered by every small cut or injury. Yet, as men get older, testosterone levels decrease and with them the threshold for and tolerance for pain. This is the physiological basis for elderly men complaining about pain which they might have disregarded when they were younger. Among women, tolerance to pain changes with age but also over the course of the menstrual cycle due to hormonal changes. In general, women are more sensitive to pain during the first half of the menstrual cycle prior to ovulation than during the second half.[6]

The immune system influences differential pain sensation between the sexes as well. The immune system of women is more active than is that of men, as is their inflammatory response. The inflammatory response is not necessarily related to infection, rather, it is a recruitment of the body's defense mechanisms in response to what the body identifies as an "attack." This response increases the chances for healing a wound or an infection, but stronger immune responses are also associated with increased risk of autoimmune diseases which are marked by a high prevalence of pain.

WHO IS MORE SENSITIVE TO PAIN?

Opinions regarding pain tolerance among men and women are contradictory. According to one view, women have a higher pain threshold and are better able to cope with pain than men. This view is based primarily on the fact that women must endure the pain of childbirth. We are familiar with the amusing statement that "if men had to give birth, the human race would soon die off." Of course, every joke has a kernel of truth to it. And indeed, during pregnancy the woman's body undergoes numerous physiological changes, one of which is an increase in pain tolerance due to a rise in progesterone level, perhaps in preparation for the anticipated pain of childbirth. Progesterone levels toward the end of pregnancy are more than one hundred times higher than before menstruation, when a moderate rise in pain tolerance is also detected.

But experimentally and clinically, women experience more pain then men, as has been observed following surgery.[7] The discrepancy here relates to information collected among hospital patients' consumption of opiate painkillers. Women's lower rate of self-administered painkiller consumption is often cited to support the general view that women are more resistant to pain than men. Researchers from Hong Kong examined data from a computerized system that enables patients to exert control over self-administering analgesic painkillers given to them by intravenous infusion.[8] The results showed that women consumed smaller amounts of opioid pain killers than did men. This could be interpreted that women respond better to opiate painkillers. Again, these findings may have contributed to the opinion that women suffer less from pain than do men. But these data can also be interpreted completely differently. It has been shown that while women do indeed consume smaller amounts of opiates, they also suffer more from the side effects, such as nausea and vomiting. Could this be the reason they choose to consume smaller amounts of these drugs when they are hospitalized? Other studies have reported precisely the opposite finding when the painkilling medications were used outside the hospital. They found that women use significantly greater amounts of painkilling medications than men, whether prescription or over-the-counter. Despite these findings, in general women are being offered less treatment for pain than are men.[9]

An accumulating body of research proves that women are not more tolerant of pain than are men. Indeed it has been shown that[10] women have a lower pain threshold and a lower tolerance for pain than men. They feel discomfort more keenly and are less responsive to treatment for pain. Moreover, we must keep in mind that most medications, including painkillers, were originally tested on men.

Here is an example from an Emergency Room. The general practice is not to administer strong painkillers in the ER, at least until a diagnosis has been made, in order to avoid masking whatever is causing the pain. This is true for male and female patients alike. Yet researchers who conducted a study in emergency rooms in the United States found that the likelihood a woman would be given painkillers for acute abdominal pain was significantly lower than for a man. Out of a thousand patients who came to the ER, 62% were given painkillers. Despite similar measures of pain for the two sexes, only 60% of the women received treatment, compared to 67% of the men. The women also received fewer opiates (47%, compared to 56% of the men) and had to wait longer for treatment (65 minutes and 45 minutes, respectively).[11] Discrimination was very unlikely the reason for this bias, since the medical and nursing staff included female and male professionals. So the reason was most related to the fact that clinical practitioners have more trouble assessing women's pain than that of men based upon preconceived gender notions. Because women in general take more medications than men and are more likely to talk about their pain, clinical practitioners are not always sympathetic to women's complaints. In contrast, they are more receptive toward men's pain, because they do not expect men to complain as much as women do. A woman is "expected to complain" about pain, whereas when a man has a similar complaint it's understood that he must **really** be in pain.

CHRONIC PAIN

Chronic pain is one of the major burdens of mankind. The National Academy of Medicine recently published a National Strategy Plan in order to address this issue.[12] Vexingly, there is a significantly greater prevalence of chronic pain among women than among men. The most common form of chronic pain is headache. One type of headache, the migraine headache, is marked by recurrent pain that is usually limited to one side of the head. In ancient Greece this illness was known as hemikrania, which means "half of the head." The condition has a genetic component, and is usually accompanied by nausea, oversensitivity to light and to noise, and can be prompted by many environmental factors. The prevalence of migraine headaches is 17% among women and only 6% among men.

The prevalence of tension headaches is also twice as high among women. This type of pain is caused by tension in the muscles of the forehead and the nape of the neck. In contrast, cluster headaches are five times more prevalent among men. This condition is a relatively rare pain that usually begins around the eye. The pain is extremely sharp, usually appears on one side of the head and is often described as the worst possible pain known to humans. Cluster headaches do not have warning signals like migraines, and the drug treatment differs for each type of pain. For this reason, the correct diagnosis is imperative.

Children are not immune to headaches, and there are gender differences among children as well. Migraines appear earlier in boys than in girls, with the peak period ranging from age five to ten, while among girls migraines appear between the ages of 12 and 17. Until puberty migraines and other types of headaches are more prevalent among boys than among girls, but their incidence and prevalence switches at sexual maturity.

Another form of chronic pain is of neuropathic origin. Neuropathic pain originates in a nerve that is injured, diseased or not functioning properly, and is twice as prevalent among women. This type of pain occurs, for example, in erysipelas disease, advanced diabetes, multiple sclerosis, various malignancies and other conditions as well.

Muscle and skeletal pain is also 30% more prevalent among women than among men. Moreover, in all areas of the body's musculoskeletal system pain is more frequent among women and they experience more associated limitations in mobility.[13] Fibromyalgia, for example, is marked by widespread musculoskeletal pain with local sensitivity in certain areas.

Other characteristics of this condition are ongoing fatigue, sleep disturbances, and bowel disruptions. Even today, no cause for this condition has been found. Scientific discussions of the risk factors for this condition have included many hypotheses, among them extreme emotional or physical trauma, long-term infectious diseases, and even genetic factors. It appears unrelated to socioeconomic status, ethnic origin, or any other attribute with the exception of gender. Almost all who suffer from this condition are women.

Another source of pain common among women originates in the digestive system. For example, irritable bowel syndrome is five times more prevalent among women than among men, as mentioned in Chapter 7. Because pain can affect many essential bodily functions, among them respiration, heart rate, blood flow, and the digestive system, the physiological system attempts to defend itself by secreting substances to reduce pain and stress. These substances, called endorphins, are neurotransmitters whose chemical structure resembles opium. Today science has identified more than twenty different types of endorphins that act in the spinal cord and the brain. In addition to their role as natural pain relievers, endorphins lead to a feeling of euphoria and enhance the body's immune reaction and even the secretion of sex hormones. Because endorphins are also secreted during continued physical effort, scientists are considering the issue of whether "addiction" to strenuous sport and the subsequent "runner's high" is related to the secretion of endorphins. It seems that even the response of this internal system activated against pain differs between the sexes, at least in experiments with hamsters. Among females the resulting secretion of endorphins was lower than among males when both sexes were exposed to the same pain-inducing stimuli.

COPING WITH PAIN

In recent years the approach to pain management has undergone fundamental changes.[14] Doctors are starting to appreciate that for chronic pain management, drug treatment does not suffice; interdisciplinary approaches and the cognitive involvement of the patient is crucial for success. Interestingly, initial pain sensation involves the classical brain regions associated with pain which have been termed "neuromatrix" and include many regions such as the cortex, hypothalamus, thalamus, and more. But over the course of chronic pain, other brain regions associated with emotional and psychosocial processing become involved.[15] This may explain why pain becomes more bearable once patients are more actively involved in treatment strategies. The goal of chronic pain treatment needs to focus not only on pain intensity and how to manage it through medication, but on the patient's ability to cope and find ways to live and function with pain. It is therefore of utmost importance that patients who suffer pain become actively involved in the management plan. It may require multidisciplinary approaches including psychosocial support, which is likely to be as effective in some cases as drug treatment. Coping with pain means to actively address issues like fear, catastrophic thinking, lack of resilience, and over-attention to pain-related phenomena. All of these tendencies are likely to draw the patient into a vicious cycle that is difficult to break with mediations alone. In a study from Spain which included 415 men and women with chronic spine pain,[16] the authors examined the role of fear-avoidance and acceptance of pain in relation to pain intensity, functional status, depression, and anxiety, comparing the two sexes. Incidentally, acceptance is not synonymous with resignation but rather indicates an active attitude. According to this study (and in contrast with previous reports) there was no gender difference in catastrophizing, defined as an exaggerated reaction to negative experience. Concerning coping with pain, there are conflicting reports. In the study discussed here, women showed significantly more anxiety related to pain and experienced more pain intensity. In spite of that they had a higher level of daily functioning. This was interpreted to mean that women might be more capable of coping with chronic pain than men. In a more recent review,[17] which was based on the assessment of seven research studies, it was reported that women in pain were more likely to use less effective coping strategies with poorer functioning, while men had better functional outcomes. So while there are no definitive findings on the topic, I'm heartened that the problem of coping with pain and its related gender issues are being deeply considered in current research.

MEDICAL TREATMENT OF PAIN

One of the most important groups of medications in the treatment of pain includes opium derivatives such as morphine. The term "morphine" is borrowed from Morpheus, the Greek god of dreams. Opioid drugs, exemplified by morphine, attach themselves to receptors located on neuronal cells in the spinal cord and the brain. In attaching themselves, they block the neural conduction path that activates the nerve cells. They then activate the nerve but at a lesser intensity, or even completely block the nerve from transmitting information to the brain. This mechanism is known as "competitive action." As is the case with other medications, pain-relieving medications operate differently in men's and women's bodies.[18] Women respond later to intravenous morphine treatment and in addition, the drug action dissipates more rapidly—after surgery for example.

Morphine and its derivatives prefer to attach themselves to particular receptors in the brain and the various opium derivatives are defined according to these preferences. Derivatives more likely to attach themselves to KAPPA receptors are known as KAPPA opioids, while others that attach themselves more to MY receptors are called MY opioids. It turns out that KAPPA opioids are more effective among females, while MY opioids are more effective among males. These findings emerged from animal studies on hamsters, but they offer insights about human beings as well. A practical conclusion from these insights is that in administering pre- and post-anesthetic pain relief, men and women should be treated differently in order to ensure the optimal anesthesia for both sexes.

Another common group of pain-relieving medications is the nonsteroidal anti-inflammatory drug group (NSAIDs). Unlike opium derivatives, these medications are not addictive, but they can cause side effects such as bleeding. Opium derivatives act mainly on the brain and the spinal cord, while NSAIDs act mainly on the peripheral nervous system, which consists of the nerves outside the brain and spinal cord. Generally speaking, these medications are more effective in men than in women, but under some conditions these drugs can be effective among women, as they reduce the production of prostaglandins. For example, prostaglandins are the local hormones that cause the contraction of the uterus during the menstrual period, and in fact menstrual pains may be connected to the increased secretion of this hormone. Often women who suffer from severe menstrual pains also experience stomachaches and diarrhea—symptoms related to the over-secretion of prostaglandins. Therefore, gynecologists whose patients complain of menstrual pains should also ask whether they have any accompanying digestive symptoms. If they do, it is likely that over-secretion of prostaglandins is causing both types of symptoms. The most appropriate and effective treatment in this case are medications that inhibit the synthesis of prostaglandins, such as Advil, Ibuprofen, and Adex.

Like all medications, pain relievers must also be absorbed so as to reach and operate at the target site. These processes are based on two mechanisms. The first is known as pharmacodynamics, or the way a drug affects the cells of the body. The second mechanism is pharmacokinetics, which involves the impact of bodily systems on drugs from the time they are taken until they are expelled. Put simply, pharmacokinetics is what the body does to the medication and pharmacodynamics is what the medication does to the body. In general, pharmacokinetics operates according to principles known as LADME, an acronym for the words describing the stages of drug action: Liberation (release of the drug); Absorption (its absorption in the body); Distribution (dissemination of the drug throughout the body tissues); Metabolism (the metabolism of the substances in the drug); and Elimination (expelling the drug from the body). Liberation is dependent only upon the drug's physical and chemical attributes, but all the other stages are strongly influenced by gender. The absorption of a drug in the body is naturally dependent upon body attributes. If absorption is through the skin, differences in skin characteristics between the sexes will have an effect on the process. If the drug is introduced into the body by intradermal or intramuscular tissue injections, the differences between these tissues in the sexes will play a role. Distribution of a drug throughout the body is dependent upon how it dissolves in water and in fatty tissue. The ratio between these two components in the body is different for the two sexes. Metabolic mechanisms also differ between the sexes, as do the efficiency and role of the mechanisms of drug elimination through the kidneys and the liver. Thus, if all the components of pharmacokinetic processes are marked by differences between men and women, there is no logic in administering pain-relieving medications, or any other medications, mindlessly in the same way to men and women.

Given this understanding, there is no doubt in my mind that pharmacy shelves will soon be stocked with different medications or different dosages for men and women without regard to their body weight but rather taking their gender into consideration. This will be true not only for pain-relieving medications but for most, if not all, medications.

The first revolutionary step in this direction was taken in January 2013 when the American Federal Drug Administration (FDA) instructed pharmaceutical companies to manufacture a common sleeping pill—Ambien—in different doses for men and women after it became clear that women need half the dosage that men do. The FDA took a second step in 2014, when it recommended that Flurazepam, also a drug for the treatment of insomnia, create different dosages for men and women. I anticipate that in the future many more medications will have recommended dosages for men and for women, that there will be drugs specifically for men and for women, and that medications will be developed with different components for men and for women. Research is heading in this direction, and with a more sophisticated understanding of pain mechanisms in the two sexes we can only hope that women, as well as men, will receive the pain treatment they so desperately need.

CHAPTER 11

TOO HOT, TOO COLD—
GENDER ASPECTS OF
TEMPERATURE REGULATION

P ROBABLY, THE MOST COMMONLY HEARD MYTH ABOUT COLD
weather is that it's liable to cause illness. Indeed, certain
viruses thrive in cold temperature, the defense mechanisms of our upper respiratory tract are restricted, and heatwave
and coldwave episodes are associated with sickness especially
among the elderly. Most common causes of illness related to
temperature changes include the heart, blood vessels, lungs and
kidneys. This is not only the case for extreme temperatures but
also for what would be considered moderate ambient heat or
cold. Women are at greater risk than men to suffer from heart
disease, sudden cardiac arrest or stroke in severe cold while in
men, heart attacks, asthma attacks and pneumonia are more
common in hot weather.[1,2]

Therefore, temperature and temperature regulation are
important in health and disease, which is the point where gender medicine comes in. Empirically, men and women experience temperature differently and also cope differently with heat
and cold. Does this have a physiological basis and if yes, how
do the sexes differ in this aspect? This chapter is devoted to
show various facets of this fascinating topic.

To begin to understand why men and women respond
differently to heat and cold, we must first understand one basic
fact: Humans are tropical animals. Although people today live
in diverse habitats ranging from the Arctic ice to the sweltering

Sahara, the general scientific consensus is that humankind originated in the equatorial regions of Africa, eventually venturing out to colonize the rest of the planet. Known as the "naked ape" for a reason, as long as we've existed we've been concerned with how to protect our bodily temperature.[3] As we left the tropics behind, our endless search for food was always accompanied by our search for shelter. Warmth equaled (relative) safety.

Because our species came of age in a hot climate, we're naturally more sensitive to cold than to heat. On our skin are four times more cold receptors than heat receptors. Interestingly, this increased sensitivity is more pronounced in women than in men.

Thermoregulation, or how the body attempts to maintain its temperature, is a complex phenomenon. Factors that affect it include age, levels of sex hormones, body composition, body mass and size, social rules, sleeping patterns, ability to sweat and more. Genetics, too, play a role in cold sensitivity. For example, people with red hair, particularly women, are more sensitive to cold than blondes or brunettes. In a strange genetic twist, it turns out that the gene responsible for red hair color is also related to body temperature regulation.

To complicate matters, there is an important difference between **being** cold and **feeling** cold. Being cold refers foremost to core body temperature, namely the temperature within the cavities of our body, including our head, abdomen, and chest. The core, also known as basal, body temperature is usually maintained at around 37 degrees Celsius and we're generally unaware if and when it fluctuates. What we feel are the effects of physiological mechanisms aimed at maintaining the core temperature. These occur predominately in our limbs and protruding body parts, like our ears and nose. In order to maintain the temperature of vital organs like the brain, heart, kidneys, and the gastrointestinal tract, our bodies divert heat away from comparatively nonessential body parts, such as the fingers, toes, and ears. The resulting decrease in temperature of these extremities makes us **feel** cold. Jumping into an ice-cold pond in the winter will certainly make you **feel** very cold. But your core temperature, for a period of time, might not change at all. On the contrary, it may even rise, because most of the blood flow to the protruding parts of our body is being restricted or has in extreme situations even been cut off. Our hands and feet will feel freezing, but the brain and heart will be toasty and warm.

In addition to automated mechanisms designed to maintain our basal body temperature, there's an adaptive reason for feeling cold. The sensation of coldness prompts us to take conscious action to conserve and produce heat. We conserve heat by reducing the surface area of our body, clenching our fists, holding our arms tightly to the body, pulling our neck between our shoulders, and becoming as compact as possible. At the other extreme, we instinctively try to dissipate excessive heat by exposing more of our body's surface area. Just watch a person lying on a hot beach with arms outstretched and hands open. (We also sweat to cool down. From a gender point of view, men sweat more easily and more intensively than women and therefore generally cope better with heat.)

Probably the single most important factor in temperature balance is the relationship between a body's exposed surface area and its volume. The smaller the exposed surface area is in relation to the volume of the body, the more efficient the temperature maintenance, both for heat and cold. A horse or a cow has a relatively small surface area in relation to its body volume and can therefore sustain both hot and cold temperatures much better than humans whose surface area is larger in relation to their body volume. Likewise, skinny people suffer more from cold weather than people with more body fat. Just consider children's vulnerability to temperature changes or how quickly rolls cool down when coming out of the oven compared to how long a loaf of bread remains warm.

Which brings us to the first notable difference between women and men regarding temperature regulation: On average, the body volume of women is about 20% less than that of men, while her body surface area is only about 18% less. In other words, women have a relatively larger surface area in compared to body volume—a clear disadvantage for temperature preservation. Just a quick clarification related to the percentages cited: The difference of 2% between body surface of women and men seems relative small in order to serve as a reasonable explanation. But if you relate this figure to the 20% decrease in body volume, then it is 2 from 20, i.e. the difference is actually 10% and not 2%.

The second difference between the sexes is the standard core temperature. It has been shown that in men this set point is about 0.3 degrees lower than in women. As described above, once the body core temperature falls below a given set point, regulating mechanisms are initiated. This means that the process begins earlier in women than in men at a given ambient temperature. Put another way, men are slower to feel the cold than women.

The third difference between men and women in coping with temperature regulation could be described as physiological strategy. Women **preserve** heat better than men, due to their having more fat within their body cavities, and men **produce** heat better than women, because they have comparatively more muscle. It's not evident to the naked eye, but if you were to compare a female and a male Olympic athlete with identical body weights, both muscular and lean with no apparent fat, the female Olympian would have approximately 15% more inner body fat than her male counterpart, and he would be endowed with approximately 20% more muscle mass. In the long run, women might be better at withstanding long periods of cold, but they would suffer more in the short term, as they are less efficient in producing heat.

Our core body temperature fluctuates throughout the day, having its low point in the early morning while peaking in the late afternoon. It also decreases when we are tired. In both conditions, women's core temperature decreases more rapidly than that of men. What this means in practical terms is that even if a woman and a man have agreed on the room temperature for the thermostat before turning out the lights at night, she might wake in the early morning feeling cold while he continues to sleep comfortably, utterly unaware of her discomfort, given that the reading on the thermostat has not changed.

Another difference in temperature-coping mechanisms between the sexes relates to hormones. In men, testosterone levels remain relatively constant and within the same range. In women, however, estrogen and progesterone fluctuate substantially throughout the menstrual cycle. For instance, following ovulation, the hormone progesterone rises and remains high until the next menstruation. The increased progesterone levels also cause an increase in the woman's core body temperature by almost 0.5 degrees Celsius. (This dual-phase hormonal pattern has been used in the past as an indicator for ovulation, and where ultrasound and hormonal measurement are not readily available, basal body temperature curves are still widely used.) For our purposes, increased basal body temperature means a higher set point for optimal temperature in women, which in turn means that they will feel cold at temperatures men perceive as comfortable. The hormonal influence on temperature regulation is evident in menopausal women as well. Hot flashes are a direct result of hormonal changes occurring during that period in life.

Social environments also have an effect on one's perception of external temperatures, and this tendency is more pronounced in women. For example, it has been found that social isolation is often accompanied by a feeling of literal coldness. Researchers from Canada placed 65 students in a closed room and divided them into two groups.[4] The first group was instructed to imagine a situation of social isolation and the second group was asked to imagine a situation of social acceptance. The students were then asked to guess the temperature of the room in which they were all seated. The first group estimated the room temperature by an average of one degree centigrade lower than the second group. Similarly, our mood affects our perception of cold. When we are in a bad mood we tend to feel colder, and when we express anger we literally warm up and even become "hot-headed." So, what does this have to do with gender? It has to do with the social acceptability of expressing emotions. In most societies, it is still more acceptable for a man to shout and express anger than it is for a woman. What this means is that the social mechanism of heating up by venting anger is less available—or at least less acceptable—for women than for men.

The differences in temperature perception between the genders has not escaped those who seek to profit from it. In 2005 the European Union introduced a norm (EN 135737) in order to standardize sleeping bags under four temperature ratings (upper limit, comfort rating, lower limit, and extreme). **Comfort rating** is defined for women as the lowest temperature at which the bag will keep an average woman warm and the **Lower limit**

rating is designed for men as the lowest temperature at which the bag will keep a man warm. These two ratings differ by almost four degrees centigrade. Consequently, leading manufacturers today produce sleeping bags cut according to men's and women's physical characteristics, with increased insulation and warmth in the hood and in the area covering the feet of women, thus conforming to the EN standards for men and women.

Understanding the gender differences in temperature regulation is not only of medical and social importance but may also have a profound economic impact. Indoor climate regulations in most office buildings are based on an empirical thermal comfort model, which was developed over 50 years ago and based on the climate needs of a typical male employee.[5] Moreover, thermal comfort levels are dependent, among other factors, on the way a person produces heat. This process is called metabolic rate, which is about 35% higher in men than in women.[6] In order to offset this higher heat production, men require lower ambient temperatures—which is why most men feel comfortable at their desk while women freeze in their offices. Humming air conditioners at open windows are therefore not a rare sight. If the thermal requirements of all workers in an office were taken into account, substantially less energy consumption would be required, with potentially tremendous savings in energy expenditures to society.

Ultimately, men and women maintain their core body temperature similarly but rely on different mechanisms to do so. Although women feel colder more often than men, their mechanisms may be more effective in the long run. Since coping with cold is apparently more demanding than coping with heat, it may be advisable for men to accommodate the needs of women when it comes to the thermostat. In other words: Let her have the remote control for the air conditioning.

CHAPTER 12

MEN—THE WEAKER SEX

A ROUND TEN YEARS AGO THE GERMAN WEEKLY MAGAZINE *Der Spiegel* published a cover story on the Y chromosome entitled "Eine Krankheit namens Mann" (A Disease Called Man).[1] Among other things, the authors of the article pointed out that throughout their lives, men, commonly thought to be the stronger sex, are more susceptible to diseases than are women, in particular to infections. This is an extremely important statement, since it indicates that size and physical strength do not translate into better health. Indeed and in spite that most research on diseases has been performed in men, they are still at a health disadvantage when compared to women throughout their life cycle, including longevity. Gender Medicine aims to improve the quality of health care in both sexes, precisely by pointing to the differences of bodily functions in men and women to the benefit of both. Therefore, to understand pitfalls in men's health serves the exact same purpose as to point to the pitfalls in women's health. The following two chapters focus on men's perspective of gender medicine.

FEEBLENESS THROUGHOUT LIFE

In general, women outlive men towards the end of life as well. Until the mid-19th century, life expectancy was less than forty years and was about the same for the two sexes. Since then it has gradually risen and a difference has emerged in favor of females (Table 1). In the Western world, the gap in life expectancy between the sexes continues to increase throughout life, so that by age 65 the ratio of men to women is 40% to 60%, and after the age of 85 it decreases to 30% men in the population as compared to 70% women. Among those who have reached the age of 100 there are fewer than 20% men and correspondingly over 80% women. Worldwide, life expectancy is higher in women than in men and in most Western countries this difference is evident for all age groups (Figure 1).[2] In all age groups, then, women have a survival advantage over men.

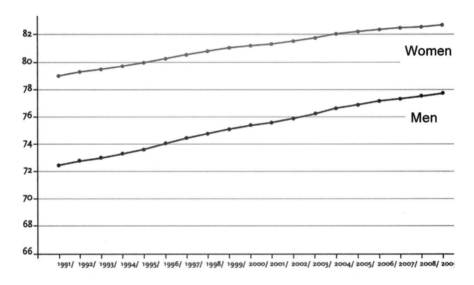

Figure 1: Mean life expectancy in Germany at birth 1991/1993

The question, then, is whether this gap is due to a longer life expectancy among women or a shorter life expectancy among men. To begin to unravel this question we must examine both genetic and environmental factors. To assess the effect of the environment on life expectancy between the sexes, researchers examined data for men and women living in Germany under the same conditions, charged with the same tasks and responsibilities, who never had children, and who were subject to a similar behavioral regime. An ideal environment for this type of research is the cloistered setting of a monastery or convent, and indeed such data were included in the study. Yet even under these ideal conditions researchers found a discrepancy in life expectancy, though in this case it shrank to one year in favor of women. The researchers concluded that this one year difference could be attributed to genetic factors, while the additional difference in life expectancy could be attributed to environmental factors.[3]

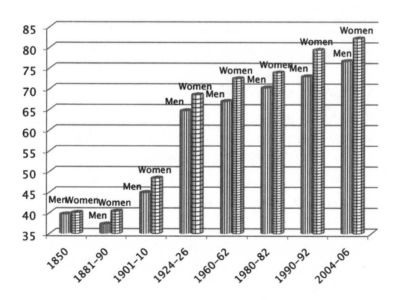

Table 1: Life expectancy in Central Europe from 1850 through 2006.

SOURCE: **Dinges**[4]

As noted, male vulnerability begins in the uterus. A study conducted at Beilinson Hospital in Israel examined the impact of the sex of the fetus on the outcome of pregnancy among 66,000 normal pregnancies with a single fetus.[5] The research revealed that while delayed growth and breech presentation (when the baby faces away from the birth canal) were more prevalent among female fetuses, women carrying male fetuses faced more challenges: premature births, premature rupture of amniotic membranes, instrumental deliveries by forceps or vacuum, a higher incidence of irregular fetal heart activity, and more Caesarean sections. The different hormonal environment in which female and male fetuses develop may play a part in this discrepancy (see Chapters 3 and 4).

In another study which sought to determine whether twins who developed in the same uterus would present different pregnancy outcomes for males and females the same authors examined the birth outcomes of 2700 pairs of non-identical twins.[6] The results showed that among pairs of male twins there was a greater prevalence of premature birth and low birth weight than among pairs of female twins. In general, pregnancy outcome was significantly better for female twins than for male twins. For mixed pairs the female had a positive influence on the condition of her male brother, while the male had a negative influence on his sister. The authors hypothesized that testosterone from the amniotic sac of the male twin passes into the amniotic sac of the female twin and has a negative impact on her, while the very presence of the female fetus in the uterus protects against some of the negative effects observed in pregnancies with a single male fetus.

Most serious complications around the time of birth occur with male fetuses. Boys exhibit more limb malformations, more cases of cerebral palsy and brain damage and twice as many cases of perinatal death before birth and in the two-week period following birth. During infancy, twice as many boys die as girls, more male infants require respiratory therapy and NICU hospitalization, and two-thirds of the infants born with immature lungs are boys. Here, too, the higher levels of testosterone in male fetuses may be the culprit.

FEEBLENESS TO ADOLESCENCE

During childhood boys appear more vulnerable than girls to numerous diseases. Malignant diseases, for example, are 25% more prevalent among boys than among girls. Obesity is also more common in boys. According to data from Germany, 10% of ten-year-old boys suffer from obesity, compared to 5% of girls, and this percentage becomes greater among adolescents and young adults.[7] And if this were not enough, the emotional world of boys points to a higher degree of vulnerability. A study conducted by researchers in the United States examined the responses of six- to eight-year-old children to the sounds of babies crying, which they heard through an audio system that included a microphone.[8] The children were given the option to either try and talk to the baby or to switch off the loudspeaker. More girls went up to the microphone and attempted to comfort the babies. What about the boys? Most of them simply turned off the microphone, twice as often as did the girls. This observation ostensibly shows that boys are less sensitive than girls. But when the researchers examined the children's heart rate variability as a measure of stress they discovered that the crying sounds of a baby were more stressful to boys than to the girls. It seems the girls, in showing empathy, found a way to express their emotions practically. The boys, in contrast, did not manage to find an appropriate way to express their emotions, so they turned to a more "radical" solution by neutralizing the source of the crying. This dissonance between emotional involvement and the inability to cope with it is quite typical of the world in which children and adolescent males live.

Unlike girls, boys—particularly in adolescence—generally find it difficult to share their emotional distress with others. One of the reasons for this is the behavioral expectations that most societies impose on boys. Most teenage boys are taught to "act like men", to not show weakness or cry. Furthermore, alexithymia—the inability to recognize emotions and to express feelings with words—is more prevalent among boys. This condition apparently has an anatomical cause related to the connectivity between the two hemispheres of the brain, which is less developed among boys than among girls.

This dissonance between emotional stress and boys' inability to express it, accompanied by societal taboos, makes things even more harder for adolescent boys, who are often "ashamed to feel shame."[9] The greater susceptibility of boys to injury, particularly during adolescence, is also related to behavioral differences. Boys make more mistakes than girls in assessing risks, tend to take more of them due to social pressure, and are generally more willing to try dangerous activities. All of these, combined with the fact that boys are generally more competitive than girls, may explain why boys are more prone to accidents, sometimes even fatal ones.

The fact that adolescent boys are more impulsive and take more risks may have an anatomical-developmental explanation as well. Certain regions of the brain such as the frontal cortex (which among other things is responsible for controlling behavior) generally develop later, and this is particularly true among boys. Moreover, testosterone, which among other things is related to impulsive behavior, reaches high levels before the teenage male brain's control system has matured. This lack of emotional equilibrium serves as the backdrop for the emotional ups and downs experienced by teenage boys, who do not always receive the proper help under these circumstances.

Often boys get help when they **cause** problems and not necessarily when they **have** problems. This may be among the reasons why suicide is the second most frequent cause of death among adolescent boys and young men in the Western world. While suicide attempts are four times more prevalent among girls than among boys, "successful" suicide attempts leading to death are four times more prevalent among boys. A girl's suicide attempt may be more of a desperate call for help, while for a boy it is the fulfillment of a decision to end a life. The means used by boys in their suicide attempts are also more violent, like jumping from height, firearms and hanging, while girls more often use medication overdose. Overall, the most frequent cause of death in adolescent boys is accidents, which are to a large extent related to their incautious behavior. Twice as many boys are killed in accidents as are girls.

Among young men illness and mortality rates from work accidents, sports, and automobile accidents are significantly higher than among young women. High-risk professions such as soldier, police officer, firefighter, and construction worker are far more likely to be filled by men than by women. Bone fractures are three times more frequent among young men than among young women, probably also due to their greater inclination for physical risk.

THE FEEBLENESS OF THE ADULT MALE
AND THE AGING MALE

Even after adolescence, men are still more vulnerable to illness. Heart and lung diseases, heart attacks, strokes, neurological diseases, diabetes, lung cancer, and especially infectious diseases are all more prevalent among men than among women. For example, the risk of post-operative infection is twice as high among men as it is among women. Furthermore, mortality rates after surgery with septic complication is three times higher among men.[10]

In the fifth and sixth decades of life about 2% of men experience ongoing despondency and even depression though in many cases neither they nor those around them are aware of it. They may experience sleep disruptions, a lack of energy, disturbances in their sexual functioning, weight gain, and muscular weakness. Some refer to this cluster of symptoms as "male menopause" or "andropause," but the resemblance to female menopause is inaccurate and purely semantic. The word "menopause" comes from the Greek words *"menos"* (monthly) and *"pausis"* (cessation or pause) and refers to the cessation of the monthly menstrual bleeding; in the life cycle of men, there is no monthly event which "pauses." In contrast to virtual cessation in the production of female hormones by the ovary, in men the production of male hormones continues, even if to a lesser extent. (In fact, this reduction begins in a man's thirties.) Moreover, for men there is no cessation of fertility. While male fertility does drop with age, men continue to produce sperm cells throughout their lives.

What the two sexes share in this period of life is the quantitative drop in the production of sex hormones. Many women suffer to some degree from menopausal symptoms, but as discussed above, some men also suffer from symptoms of hormonal deficiency. The common term to describe this condition is ADAM, an acronym for Androgen Deficiency of the Aging Male. Another term used is SLOH, an acronym for Symptomatic Late Onset Hypogonadism. Just as women with hormonal deficiency can be treated, so too can men. Yet male patients are less aware of the signs of illness, seek out medical help less frequently, are less willing to talk about their medical problems, and tend to be less forthcoming in providing relevant information. The problem in diagnosis stems from a lack of awareness among doctors and patients as well. Moreover, many signs of ADAM are likely to be related to depression, to aging, and to other diseases such as reduced function of the thyroid gland (hypothyroidism).[11]

Because masculinity is generally defined as the ability to suffer stoically without complaint and to hide any signs of weakness, illness is considered non-masculine. Therefore the concept of masculinity, including men's self-definition may lead to a reluctance to admit medical problems, to feel "ashamed" about symptoms which relate to masculinity and therefore to delay doctor's visits. Moreover, whereas the need to treat the symptoms of menopause in women is not a matter of controversy, when it comes to men, there is no consensus regarding the need for treatment. Many studies on menopause comprising tens of thousands of participants have been conducted on women. Research regarding hormonal deficiency in men is extremely scarce and generally did not include enough participants to reach any definitive conclusions. ADAM syndrome is usually diagnosed based on symptoms and tests of male hormone levels. Based on the diagnosis, various forms of testosterone treatment can be given. Many physicians are reluctant to prescribe testosterone because of the fear of prostate cancer, which has been implicated in this type of treatment. However, the claim that testosterone treatment raises the risk of developing prostate cancer has never been proven scientifically. But this situation does not need to be accepted. The Endocrine Society published in 2010 clinical guidelines which define the various types of available treatment modes available and also list contraindications for treatment.[12] Given these guidelines, in most instances, men suffering from ADAM should not remain untreated.

So far in our discussion of Gender Medicine we have focused on underscoring the idea that research on diseases and medications has primarily been carried out on men, with the hope that women be included in medical research in the future. Yet there are a number of areas where the opposite is the case, and research has been conducted mainly on women with results that cannot easily be applied to men. In the next section I'll briefly discuss three such diseases and their implications for male health: breast cancer, depression, and osteoporosis.

BREAST CANCER IN MEN

Breast cancer in men is a rare disease constituting less than one percent of the incidence of diagnosed breast cancer each year in the Western world. Yet, given absolute numbers breast cancer in men needs further study. A multinational study that included Denmark, Finland, Norway, Sweden, Singapore, and Switzerland examined data collected over a four-year period regarding breast cancer.[13] The study found 460,000 cases of breast cancer among women and 2,700 among men. A similar ratio was also found in North America and elsewhere in the world.

As a rule, men with this disease are treated based on the knowledge and experience accumulated from treating women. However there are important factors to consider: The disease appears five to ten years later in men than it does in women and with a worse prognosis; the risk of contracting the disease does not drop after age 50 as it does in women; and the treatment offered is, as noted, based on treatment protocols developed in female patients. Some of these treatments have proven to be less effective among men. One of the reasons for this may be that more than 20% of patients do not diligently continue taking long-term medications such as Tamoxifen because the side effects are more severe among men.[14,15] The second most frequent type of treatment, aromatase inhibitor drugs, is not effective unless treatment to decrease testosterone levels is taken at the same time, and testosterone depression has side effects of its own.[16] Worst of all, because of the delay in diagnosis among men, the disease is often discovered at later stages and the prognosis is worse. Little is known about the side effects of chemotherapy and local treatment or about the emotional impact on men being treated for breast cancer. Apart from this, the genetic background and family history related to breast cancer are similar for the two sexes. Around 10% of men who have the BRCA2 gene mutation will develop breast cancer in the future, and their children are at greater risk of breast cancer.

DEPRESSION AND MENTAL ILLNESS

Depression is a highly prevalent illness. It is estimated that in the United States 16 million people are diagnosed with depression each year.[17] In Australia more than 10% of visits to family physicians are reported to be because of depression.[18] The professional literature usually states that depression is twice as prevalent among women as among men, though recently this statement has been challenged as we shall discuss later. In any event, it is a growing problem worldwide and it is important that depressed people of both sexes receive appropriate treatment and care.

203

Psychiatrists generally diagnose depression based on the appearance of such symptoms as sadness, loneliness, a drop in self-esteem, a loss of interest in one's surroundings, sleep problems, lowered or increased appetite, a lack of energy, and neglect of one's outward appearance. The PHQ-9 (Patient Health Questionnaire) commonly used to diagnose depression includes nine questions, most of which refer to these symptoms (table 2). Unfortunately for men, these symptoms are based on typical complaints of depressed women.

1. Little interest or pleasure in doing things
2. Feeling down, depressed, or hopeless
3. Trouble falling or staying asleep, or sleeping too much
4. Feeling tired or having little energy
5. Poor appetite or overeating
6. Feeling bad about oneself — or that you are a failure or have let yourself or your family down
7. Trouble concentrating on things, such as reading the newspaper or watching television
8. Moving or speaking very slowly. Or the opposite — being fidgety or restless and moving around a lot more than usual
9. Thoughts that one would be better off dead or of hurting oneself in some way

**Table 2: PHQ self-administered questionnaire
for diagnosing depression**

Among men, depression can present not as passivity or lack of energy but as aggression, alcohol and drug abuse, or exaggerated interest in sex or in work.[19] Because the PHQ and other such questionnaires do not make reference to any of these symptoms, depression among men is likely to go undiagnosed.

Attention to this problem has increased, however—in recent years new methods have begun to be developed for diagnosing depression in men. Researchers in the United States examined the prevalence of depression among close to 7,000 men and women and recently published the results in the *Journal of the American Medical Association (JAMA Psychiatry)*.[20] It seems that, given appropriate diagnostic tools, the prevalence of depression does not differ between the two sexes. There is a clear tendency to stigmatize male depression.[21] A woman may be able to turn to a friend and complain about her mood or even her mental state, but this type of sharing is far less common among men. Complaints of depression by men are seen as shameful, as signs of weakness. Even as children we learned that "boys don't cry." In Australia a survey conducted among 1,100 general physicians asked about the diagnosis of depression in men.[22] Of the respondents, 64% reported having difficulty arriving at a diagnosis. This difficulty was even greater among female physicians, 73% of whom reported having more trouble communicating with male patients. The survey also showed that men are hesitant to describe their problems to female physicians. In most cases they only come for medical consultation in the midst of a crisis, while women tend to seek help when they feel a crisis coming on.

Even postpartum depression, which occurs in around 10% of women, is not exclusive to women. Depression before and after delivery may be present in a similar magnitude among men.[23] And just as a mother's postpartum depression raises the risk of a child developing emotional disturbances in the future, the same is true if the father becomes depressed.[24] Another important reason why correct and reliable diagnosis of depression in men is crucial.

OSTEOPOROSIS IN MEN AND IN WOMEN

Osteoporosis is defined as a bone disease characterized by a loss of bone mass and a change in bone quality causing increased skeletal fragility.[25] The process of decreased bone density often begins in the thirties and has no noticeable clinical signs. The danger of osteoporosis is not the loss of bone mass itself but rather the growing risk of fractures, particularly of the hip.

Historically osteoporosis has been considered a disease of women and indeed its prevalence rises sharply after menopause, arriving with the decrease in levels of estrogen. In general, bone density decreases in women at a rate three times faster than in men. One in every two women is likely to develop this condition during her lifetime. Yet, while this is also the fate of one in five men, most people are not aware that men can suffer from osteoporosis as well.

One in six women is at risk of breaking a hip over the course of her lifetime, compared to the one in nine risk of developing breast cancer.[26] And although it's more likely that a woman will break a hip than be diagnosed with breast cancer, popular attention has skewed our attention and resources to the latter. There have been major advances in the treatment of breast cancer, both in preventative and diagnostic care. With respect to the treatment of hip fractures, however, the situation is less encouraging—mortality within one year of diagnosis is approximately 30%.

It is estimated that worldwide, nine million people each year suffer from fractures caused by loss of bone density, 30% of whom are men.[27] Even though this condition is more prevalent among women, illness and mortality from hip fractures are more frequent among men. This could be because men are less likely to be diagnosed or referred for treatment for osteoporosis.[28]

Bone strength is a combination of density and hardness. Bone density stems from the mineral content of the bone, while hardness is dependent upon bone structure. Here too there are many differences between men and women. Bones consist mainly of dense areas which constitute around 80% of the skeleton and enclose sponge-like tissues containing the bone marrow. These spongy tissues help reinforce the structure and ensure that the bone is not too heavy (Figure 1a).

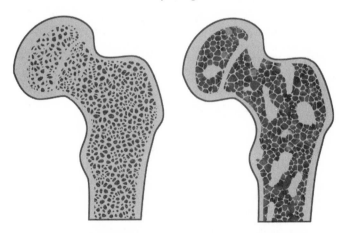

Figure 1a: Healthy bone   Figure 1b: Osteoporosis

Over the course of osteoporosis in women, the connections between the sponge-like tissues are torn. In men these connections grow thinner but preserve their existing structure, thus maintaining more effectively the continuity and stability of the bone tissue. In women, however, these tears disrupt continuity and make the bone more fragile.[29] (Figure 1b)

The prevalence of osteoporosis among men is 12% and among women it is 29%. However, I suspect that the real difference may be smaller, with relatively more men actually suffering from osteoporosis. The current definition of osteoporosis for postmenopausal women and men aged 50 years or more was set by the World Health Organization (WHO) and is based on the mean bone density of the hip bone in Caucasian women between 20–29 years of age.[30] For both men and women, then, the reference standard is a young white female. Considering that the bone mass and density of young men are greater than those of young women, the measured difference between older men and younger men will in effect be larger than that measured today according to the organization's recommendations. Another important fact to take into consideration: while bone fractures are more common among women, as mentioned above, men are more likely to grow ill or die from any fractures. A study from Canada that included close to 4,000 patients with hip fractures included 29% men compared to 71% women. Over the course of hospitalization, however, the death rate among women was 4.7% while among men it was 10.2%. Even though hip fractures are three times more frequent among women, the mortality from these fractures is twice as high among men.[31]

Generally, screening for osteoporosis is recommended by most professional societies worldwide for women who are 65 years and older and for men 70 years and older. Virtually all professional societies have reached a consensus on screening for women, but not for men. Surprisingly, the US preventive services task force did not find sufficient evidence to issue a recommendation for screening in men.[32] Consequently approximately four times more women undergo bone density testing than men. Between 1999 and 2011 around 41,000 people were referred for bone density testing to the Rabin Medical Center. Only 11.5% of these were men.[33]

In summary, even though men are stronger and larger physically than women, they are surprisingly vulnerable during the entire course of their lives. Men have sometimes been described as the "extreme gender."[34] A small fraction of the male population dominates in most echelons of modern society but on the opposite side of the coin, a large fraction of the male population is vastly overrepresented among criminals, abusers of drugs and alcohol, the homeless, people who commit suicide, and in cases where things go wrong in life. It seems men are less well equipped to cope with adverse situations like life crises, divorce, financial difficulties, and disease. This is mostly due to the gender role men have assumed in society but also to an intrinsic biological vulnerability as discussed above. Moreover, as women are entitled to being treated for diseases which affect them based on research which has been performed in women, men are entitled to a similar approach. The gap to be closed is still far larger for women but there is also need to close the gap for men.

CHAPTER 13

ARE MEN AN ENDANGERED SPECIES?

W HILE THE TITLE OF THIS CHAPTER MAY SEEM CHEEKY TO some, there are scientists who have seriously suggested that the male of the species will disappear in a matter of a couple of hundred thousand years. Before we get to that, however, let's look more closely at male physical vulnerability and feebleness. Why are men so susceptible to disease and infection? The main reasons are chromosomal and hormonal, and I'll start with the primary culprit: the male hormone, testosterone.

Testosterone is among others responsible for suppressing the immune system, which protects us from infectious diseases. This explains why men are less protected from infectious diseases than women. An experiment with male hamsters revealed that shock as a result of massive hemorrhage caused only a moderate immune response.[1] Their immune response improved when testosterone was lacking. Female hamsters demonstrated a better immune response from the outset, but the administration of testosterone lowered the intensity of their response.

Testosterone also constitutes a risk factor for cardiovascular diseases—it causes a reduction in "good" cholesterol and a rise in "bad" cholesterol. As a result men have a greater tendency to develop plaque in their arteries and experience arterial narrowing. This is particularly dangerous for the coronary arteries that supply the heart muscle, and explains why heart attacks and strokes are more prevalent among men under the age of fifty compared to women in the same age bracket. Women in this age group are protected by estrogen, which lowers the frequency of heart attack and stroke (see Chapter 7). Testosterone is also related to risk-taking behavior at work, during leisure activities, and driving, as well as to excessive drinking, all of which contribute to increased risk of injury and death. The masculine hormone is both a blessing and a curse. It makes men the stronger sex physically, but makes them the weaker sex over a lifetime.

However, it's not only hormones that affect male weakness. Genetic factors come into play as well, including stem cells and the deterioration of the Y chromosome.

GENETIC FACTORS FOR MALE WEAKNESS

After the sperm cell and the egg merge in successful fertilization, the resulting cell contains all the genetic material of both parents. This stem cell is totipotent—in other words, it has the potential to develop into any cell in the human body. The early embryo begins to divide, and up to four days after fertilization when the embryo already contains dozens of cells, each newly created cell is still totipotent. If at this stage the collection of cells divides, each group can develop into a complete organism, as in the case of identical twins.

At a certain stage, when the dividing cells begin the process of differentiation, the fate of individual stem cells is set—the type of cells into which they will develop is determined. Even though each of the cells still has the same genetic makeup, from that point forward the stem cells will henceforth use only part of their gene pool. The genes used by each cell enable it to perform its function and develop into a lung cell, or a heart cell, while the genes it does not need are "turned off" forever. Such a cell could not spontaneously turn into, for instance, a liver cell. This is also how the gametes—sperm cells and eggs—are created. All the other stem cells are directed at generating the cells of the various body systems and are known as somatic cells.

The genes are located on the chromosomes and provide hereditary information in the form of a "recipe for action." Each chromosome is a long molecule consisting of genes arranged in sequence along two long helixes that are intertwined with one another in a double spiral. The researchers James Watson, Maurice Wilkins, and Francis Crick were awarded a Nobel Prize in 1962 for discovering this structure. The number, size, and shape of chromosomes differ among animal and plant species. Each human cell, except for sperm cells and eggs, contains 22 pairs of somatic (body cell) chromosomes (which are also called autosomes) and two sex chromosomes, adding up to 46 chromosomes in each bodily cell. Half of these are from the mother and the other half from the father. All pairs of autosomes are similar in shape, size, and sequence of their genes. The sperm and the egg are unique, however. The sex chromosomes are different in size and shape. Each of these sex cells contains only half of the genetic set, that is only 23 chromosomes, so that the next generation will inherit genetic characteristics from both parents and will have the proper number of 46 chromosomes after the sperm and egg merge during fertilization.

A human's genetic sex or genotype and chromosomal structure are determined by the combination of sex chromosomes. Women and men both have 22 pairs of somatic chromosomes, 44 in total, and two sex chromosomes. In women these sex chromosomes are XX and in men they are XY. An egg always has one X chromosome but sperm cells may carry either an X or a Y. A woman's genotype is therefore 46 XX and man's genotype is 46 XY.

The X chromosome and the paternal Y chromosome also differ greatly in their structure and genetic makeup. When the sperm cell merges with the egg during fertilization and also when body cells divide throughout life, all 23 pairs of chromosomes from the male and the female align themselves facing one another with similar genes adjacent to one another. In most cases one of the facing genes will take precedence over the corresponding gene on the opposite chromosome. Take, for instance, the gene that determines hair color. Let us assume that the gene from the mother dictates that the offspring will have light brown hair and the gene from the father dictates that its hair will be black. The child will not be born with an average of the two genes. With the help of complex mechanisms, the expression of one of the genes will be suppressed and the corresponding active gene will determine the child's hair color. A similar process occurs throughout life for another purpose: to correct defects in stem cells designated for various organs. The gene that is not expressed serves as spare parts for the corresponding active gene on the parallel chromosome.

This spare parts warehouse is essential because hundreds of thousands of molecular defects occur in cells on a daily basis. Most of these do not require repair, but for some, repair is essential. Several mechanisms exist for this purpose. One important method is related to the alignment of identical genes located on the two chromosomes. When a defect needs correcting, the inactive gene aligned to the active gene can be used as a template to repair the corresponding gene, and the newly created segment replaces the damaged segment (Figure 2). This process is possible only if the damaged and the intact gene are structurally identical and can be physically aligned. This is the case in women, where all 23 chromosomal pairs are identical, including the X sex chromosome.

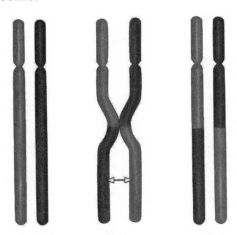

Figure 2: Repair of defects by exchange of segments between two somatic chromosomes (Crossover).

So what happens, then, in men? While the 22 pairs of somatic chromosomes are identical and can activate this corrective mechanism, the sex chromosomes are an exception. The Y chromosome does not have a corresponding chromosome that is similar in size and shape. The X chromosome is twice its size and resembles it neither in structure nor in genetic makeup. Researchers have found that three million years ago, at the outset of the evolution of mammals, the two sex chromosomes resembled each other in size and in number of genes. But lacking an effective repair mechanism, the Y chromosome began to deteriorate and has already lost two-thirds of its size and 90% of its genetic content. The female X chromosome carries approximately one thousand genes, while the male Y chromosome carries only around eighty genes, of which large regions are a genetic wasteland. Twenty-five percent of the cases of male infertility are related to genetic defects of the Y chromosome. (Among the seemingly unnecessary genes preserved on the Y chromosome is one responsible for the growth of nose and ear hair in aging men.) Yet, one of the most important genes in men is also located on the Y chromosome: the SRY gene responsible for the development of the testicles (see Chapter 3).

Chromosomes are passed down from one generation to the next, and this is true for the sex chromosomes as well. An X chromosome from the father will always find another X chromosome from the mother so that the described corrective mechanism can be activated. The Y chromosome, on the other hand is on a "dead-end alley" because under normal circumstances it will never meet another Y chromosome. No healthy males carry two Y chromosomes and fertilization leading to males with a chromosomal pattern like XYY will result in males who are infertile and are burdened with a variety of problems. Therefore, nature has created alternative repair mechanisms for the male sex chromosome, which are less effective but quite interesting.

Because of the differences in the structure of the sex chromosomes, it is impossible to line up parallel X and Y chromosomes. Accordingly their capacity for mutual repair is limited. Only when the X chromosome is folded can its ends touch the ends of the Y chromosome, so that genetic mixing can occur at the ends (Figure 3a). Defects at the center region of the Y chromosome, however, cannot be corrected with this method. As a result, over the course of millions of years of evolution the Y chromosome has developed amazing capabilities for self-repair. In a desperate search for an identical genetic region to connect itself to, it folds in upon itself and attaches two regions of its own "body" to one another (Figure 3b). In order that the genes in these two regions will indeed be aligned parallel, the genetic material there is written like a palindrome.

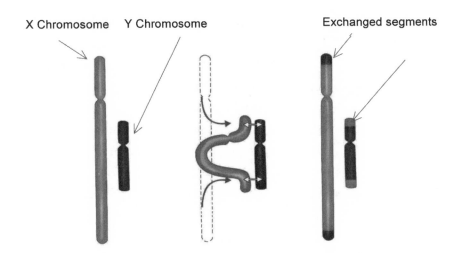

Figure 3a: Genetic crossover and chromosomal repair between X and Y chromosome

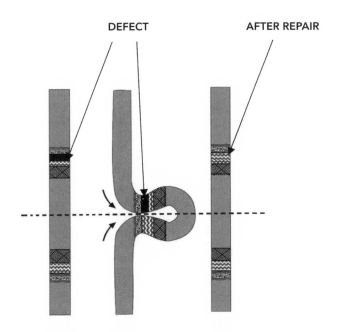

Figure 3b: Self-repair of the Y chromosome by means of folding the body and aligning similar segments to each other

That is, it can be read identically both backward and forward, like the words "madam" or "civic." One part of the body of the Y chromosome serves as spare parts for another part. But even these sophisticated techniques provide a partial solution and then only for genes located at the ends of the Y chromosome and in a small region at its center. As a result most of the genes on the Y chromosome lack a repair mechanism.

The outlook, therefore, is rather dire for the long term. Geneticists have calculated that if the process of Y chromosome deterioration continues at the same rate, in another two hundred thousand years, approximately, the chromosome will cease to exist, and *the male sex as we know it today will disappear as well.*[2, 3]

CHAPTER 14

GENDER ASPECTS OF DOCTOR–PATIENT RELATIONSHIPS

M EDICINE HAS ADVANCED AT A BREATHTAKING PACE DUE TO scientific discoveries, technological advancement, the information revolution, and the continuous specialization of medical disciplines. Such developments have fundamentally affected the organization of medical care, especially in hospitals. The latter have become medical centers concentrating on teaching and research. The results, in terms of improving health and saving and prolonging lives, have been dramatic. Yet also changed in this course of development is the traditional doctor–patient relationship. Today, many patients perceive the health care system as an overpowering and amorphous structure with countless anonymous professionals who seem to care much more about the disease of the patient than about the patient as a person.

While private practitioners and community clinics still maintain a certain intimacy, most hospitalized patients today miss the continuity of medical care, as they are not free to choose their physician. During a stay at the hospital, they will be examined by numerous health care givers who will only address them by their name after referring to the chart. The encounter with the physician will concentrate mainly on "the essentials," which today means taking the patient's history, noting medical complaints, and physical examination. Moreover, the digital age and its health care apps have also greatly changed the medical encounter. As of this writing, more than two hundred thousand smartphone applications which deal with lifestyle and health are available. Worldwide this is a ten-billion dollar industry. A recent survey from Germany has shown that 30% of patients would welcome online encounters with their physicians, 16% believe that an app could replace a visit to a doctor, and 85% of patients would be willing to load their medical file on a remote server.[1]

This new medical environment has created fundamental changes in doctor–patient relationships. Physicians are now referred to as "health care givers" and patients as "customers," which illustrates just how much the medical ecosystem has become affected by market conditions, where customers shop for services and where competition between health care providers is on the rise.

Expectations have changed as well. The trust-based personal encounter has given way to a technology-driven diagnostic and therapeutic system in which patients are required to invest trust in a system rather than in a person. Yet they expect to get the best possible results. Anything less than treatment success, with absence of failure as a given, is interpreted as negligence— malpractice suits have become a huge and threatening cloud hovering over medical care in the western world. As a result, an increasing number of physicians, in order to avoid such lawsuits, are turning to a defensive practice of medicine, performing unnecessary (and often expensive) tests and procedures to appease the patient and preempt claims of negligence. Insurance premiums for physicians are skyrocketing, and medical care continues to become more and more expensive. A multi-billion dollar industry of consultants and lawyers has emerged, feeding off this clash between patients and health care providers.

This environment exceeds by far the legitimate and necessary monitoring of medical care quality and the rightful castigation of medical negligence. Nor has it done much to improve the system or the doctor–patient relationship. Often overlooked is the importance of good communication between doctor and patient. Although poor communication is virtually never listed as a main claim, it is often the underlying motivation that causes a patient to sue. Physicians will be sued more often for dysfunctional communication than for actual medical failures.[2]

Malpractice suits are an increasingly common way for patients to punish their physicians for inadequate communication. Apart from the continuous effort to improve quality in medicine, the single best tool to avoid or reduce litigation is honest and transparent communication. The University of Michigan Health Care System introduced some years ago a disclosure program based on transparent communication with patients concerning medical errors. Although the number of medical interventions increased during the time period examined, the number of malpractice suits decreased significantly. Moreover, time required for claim resolutions was substantially shorter. Patients seem to accept medical failures as long as communication with their health care provider is satisfactory. This study clearly indicates that transparent and honest communication with patients is a key element for preventing litigation.

More importantly, it is a key issue in any doctor–patient encounter.

MODELS OF DOCTOR–PATIENT
ENCOUNTER AND COMMUNICATION

In the not so distant past (and even today in some rural communities), family physicians knew their patients and the patients' families rather well. Physicians were aware of the goings-on of their community, and the home environment of a particular patient and his or her personal or social problems. The relationship between doctor and patient was characterized by longevity and a paternalistic approach. Nothing less than blind trust was required from the patient. On the other hand, patients expected from their physician loyalty, empathy, professional knowledge, and unwavering efforts on their behalf. Once patients were convinced that their doctors had done everything in their power, they were prepared to accept failure.

Of course we tend to idealize the past and forget that physicians then were typically unaware of medical ethics as we understand it today. Topics such as patient's autonomy, right of choice, and right of information were virtually absent in the doctor–patient relationship. Physicians were guided mainly by the maxim: "First do no harm." And more often than not, refraining from harm and showing compassion were all doctors could offer to patients. Then again, listening to a patient; showing compassion and communicating appropriately; and knowing and relating to the patient's values are by themselves powerful therapeutic tools.

One of the modern scientific methods to assess the effectiveness of a drug is a clinical trial, in which patients randomly receive either the drug to be tested or an identical-looking drug, which has no active ingredients and is called a placebo (in Latin, "I shall please"). Neither physicians nor patients are aware which patient receives the drug, and which receives a placebo. At the end of the clinical trial the data is "unmasked" and the effect of the drug compared to the placebo. It has consistently been demonstrated that patients who received the placebo show a beneficial effect rate of about 30%.[3] This so-called placebo effect is a direct result of the patient's belief that he or she is receiving an effective treatment. Many of the treatment successes of earlier physicians may have been based on the placebo effect, achieved not by advanced medicine but by a trusting relationship and effective communication.

This model of the paternalistic village doctor, so often described in medical literature, has largely been replaced by that of the "informative–advisory relationship." In this model, the doctor provides all available information to the patient about his or her medical situation, gives professional advice on the offered alternatives and allows the patient to decide which course to take. The onus of decision-making is transferred to the patient. This problem becomes even more obvious if a patient is confronted with a second opinion that contradicts the first doctor's advice. All roads invariably lead back to the need for a patient's trust that the physician will offer the best possible help. A recent development along this vein is the model of Patient-Centered Care. The Institute of Medicine defines Patient-Centered Care as: "Providing care that is respectful of and responsive to individual patient preferences, needs, and values, and ensuring that patient values guide all clinical decisions."[4] This basically means that the decision-making process should be based not only on technical aspects, such as test results and an examination of the patient, but should also involve active engagement of patients and families at each level. This approach has led to the formulation of the five principles of Patient-Centered Healthcare, namely 1. Respect, 2. Choice and Empowerment, 3. Patient involvement in health policy, 4. Access/support and

5. Information.[5] Much public effort is currently being invested in promoting this approach.[6]

• • •

Regardless of which you choose, the common denominator for each model of direct doctor–patient encounter is communication. This is where patient care begins and where the stage is set for subsequent interaction. Appropriate communication allows trust to build, and ensures that all necessary information is shared between physician and patient. Appropriate communication also improves treatment results. Unfortunately, medical schools still place relatively little emphasis on teaching communications skills, focusing instead on methods and skills to obtain as much as possible factual information relevant to making the correct diagnosis. We often forget that behind the care giver and the patient are human beings with his or her own value systems, beliefs, anxieties, and expectations. Tools for communication, like verbal and nonverbal language apply as much to the doctor–physician encounter as they do for encounters between people in other life situations.

Given the inherently unequal power dynamic, the physician needs to take the lead in building a relationship founded on trust and receiving reliable information as the basis for designing an individualized diagnostic and therapeutic plan. The patient should be able to receive, understand, and later recall this information and guidance. In the unwritten covenant, both physician and patient are expected to provide truthful and easily understandable information. Finally, enough time should be set aside for documentation, writing instructions, signing consent forms, and more. The patient should leave the encounter sufficiently satisfied that she or he can follow the instructions given. All this has to occur within a rather short time frame of not more than 20–25 minutes, usually even less.

Challenging as such an ideal seems, we haven't even touched upon the other major challenge of communication in the doctor–patient relationship: truthfulness. In a recent survey, conducted by a medical website and led by Dr. Caplan, chair of the department of medical ethics at the University of Pennsylvania, on nearly 1,500 men and women, 13% of patients reported that they had lied to their physician and an additional 32% admitted to having stretched the truth.[7] Approximately a third of patients lied about adherence to the physician's instructions and to exercise, 20% about smoking and sex, and 6–7% lied about family and personal history and about getting a second opinion. About 50% of patients explained that they had lied because they feared judgment and 20–30% because of embarrassment or because they felt the doctor would not understand. Patients may lie because they do not want to have a specific piece of information in their record, because they perceive the topic to be taboo, because they fear consequences, or because they expect a fringe benefit, like securing an earlier appointment or a desired opinion letter by exaggerating symptoms.

Nor do physicians always tell the truth. Sometimes they find it too difficult to convey a dire message concerning the patient's health, sometimes they are asked by the patient's family to conceal the truth, and sometimes doctors simply don't think that the patient needs to know certain facts. In a survey performed at Harvard in the USA including close to 1,500 physicians, over 10% of physicians admitted that they had lied to patients, nearly 20% refrained from disclosing medical errors for fear of being sued, and over 55% of physicians reported that they had provided an exaggerated prognosis of a treatment.[8] Concerning transfer of information: Patients are often not given a chance to complete their opening questions. In approximately 70% of visits, family physicians interrupt the patients after less than 25 seconds and redirect their question. Less than 30% of patients complete their original statement during the visit.[9] Clearly, these examples illustrate patterns of deficient and often harmful communication.

Given these challenges, how can we cope? How can we improve doctor–patient communication so that patients receive and understand the treatment that they need? The key may be to first understand how effective communication of any kind works (or fails). The rest of this chapter is devoted to a discussion of more general issues, such as communication channels between people in general, communication characteristics in men and women, and between same-sex or different-sex pairs in particular. Eventually I'll examine to what extent the dynamics in a dyad (pair) situation apply to the doctor–patient relationship. Gender Medicine is interested in the outcome of medical care in women and men, and the doctor–patient encounter in all of its aspects plays the central role in this endeavor.

HOW DO PEOPLE COMMUNICATE?

Communication between humans is a complex endeavor. We communicate verbally and nonverbally and within these major channels we also rely on a plethora of sub-channels. We modify our verbal communication using nonverbal tools such as intonation, the volume of our voice, pitch, pauses, and emphasis. Words become meaningful in a specific cultural or social context or sometimes due to a certain intonation. The same is true for nonverbal communication. Body positions, gestures, posture, or the way we touch ourselves during a conversation may have different meanings within different contexts. For example, scratching your forehead may mean that you are thinking, stressed out, or may just mean that your forehead is itchy.

Crossing your arms in front of your chest may indicate aggression, defense or may simply be a sign that you are feeling cold. Moreover, in different cultures the same nonverbal signals may have a completely different meaning. Sitting with your legs crossed would be perceived as rude in Turkey but as a sign of being at ease when in North America. Suppressing or exaggerating facial expressions may convey different messages in different cultures, for instance searching or avoiding eye contact.

Our arsenal of nonverbal channels is even more complex than the tools that we use to modify our verbal messages and includes as major "supra channels" facial expression and body language. Facial expression is probably the most important of these supra channels. To recognize and analyze facial expression has always been crucial for social life and protection and many of our brain resources are devoted to optimizing these capabilities. Specific brain areas are reserved for recognition of faces and of facial expressions though not for any other object. As a result, humans are able to determine a person's emotional state within a fraction of a second.

Finally, nonverbal communication occurs largely in a non-intentional way, meaning that we are typically unaware of the messages and signals we send via body language. We are more effective at controlling some aspects of nonverbal expression, such as facial expression. Though we may not know it, nonverbal communication may reinforce or contradict the words coming out of our mouths, rendering us deceptive or earnest. Deciphering human communication includes the ability to interpret nonverbal and verbal messages and to compare their content.

In every conversation, people use a combination of verbal and nonverbal communication and are to a certain extent free to choose which combination of channels to use for specific inter-human encounters or for conveying specific messages. We also use "external" nonverbal channels like the way we dress for an encounter, the environment which we choose for a meeting and innumerable clues and messages which we introduce or use in our surroundings to get a message across. People choose the communication channels that seem the most suitable for the message they intend to deliver. If you want to convey a straightforward message avoiding nonverbal interference as much as possible, you may want to restrict your communication to the verbal channel and therefore prefer a phone conversation. Or you may want to further restrict information transfer to the bare essentials and avoid or omit nonverbal components of speech, like intonation, pauses, volume and therefore choose to send your message by e-mail or even snail mail. On the other hand, if you are negotiating you may want to maximize the available dimensions of communication. In such circumstances, you would want to read your opponent and would therefore insist on a personal encounter, choosing time, location, and even seating arrangements carefully. Likewise, it would be inconceivable to e-mail a patient to inform him or her about the result of a biopsy which showed malignancy. On the other hand, it is not necessary to call a patient in order to inform him or her about an appointment, as an e-mail will do as well. Unfortunately, due to lack of time and communication skills, all too often doctors choose the wrong channel for conveying these type of messages, leading to the erosion of trust in the doctor–patient relationship and the growing belief that the personal touch in medicine has been lost.

DO WOMEN AND MEN
COMMUNICATE DIFFERENTLY?

Numerous articles, papers and books have been written about how men and women use language differently. This should not come as a surprise. In most parts of the world, boys and girls are brought up with different norms and value systems, taught to play gender-specific games, engage in different types of conversation, and not always use conversations for the same purpose. Verbalization capacity also differs between men and women (see chapter 3).

(Before we go on I would like to stress again a point applicable throughout this book: whenever I mention differences between men and women, I'm not referring to **all** women nor to **all** men, nor is it the intent to promote stereotypes about typical male and female behavior. When pointing to differences, I merely conclude from a vast body of published research that certain phenomena are more common in women or more common in men.)

Just as language comprehension is intimately related to the cultural background of an individual, an individual's gender also affects his or her use and understanding of language. In fact, the renowned linguist Deborah Tannen, coined the term *genderlect* to describe the specific way language is tied to gender. She states that *"communication between men and women can be like cross-cultural communication, prey to a clash of conversational styles."*[10] In other words, both sexes use the same words and grammatical rules but gender may be an important modulator that gives these communication channels different meanings. Literature provides ample examples for genderlect. Look at the dialogues in the works of Flaubert's *Madame Bovary*, in Shakespeare's *Hamlet*, in Tolstoy's *Anna*

229

Karenina, in Oz's *My Michael* (see footnote for an example)*
and in the works of other writers from the ancient to modern
times, genderlect has always been there. George Bernard Shaw
once said that "England and America are two countries sepa-
rated by a common language." It sometimes feels as though
this could apply to the two sexes, as well.

Women are more verbally and nonverbally expressive.
During the course of a conversation, women smile, touch
themselves, use their hands expressively and orient their bodies
toward the person they're speaking to more than men do. They
are also better at decoding nonverbal communication than men
and may even rely more on nonverbal communication than the
spoken word.[11] Specifically, women perform better than men
when asked to decipher facial expression (see also chapter 1).
This skill is probably genetically hardwired, since raising infants
requires mothers to rely in particular on the ability to decode
the facial expressions of their offspring.

Women are therefore uniquely skilled at deciphering a
nonverbal message to assess the veracity of the spoken message.
This is probably what is meant in common parlance by "female
intuition."[12] We would expect women to be less prone to be
fooled in general and especially with respect to facial expressions. Apparently this holds in such situations as interviews but
less so when it comes to sales-related negotiations. In fact, due

---

*Dialogue from *My Michael* by Amos Oz. Translated into English by Nicholas
de Lange in collaboration with the author. Borzoi Books/Alfred A. Knopf Inc.
New York, NY 1972:

"Do you find me ugly, Michael?"

"You are very precious to me, Hannah."

"If you don't find me ugly, why don't you hold me?"

"Because if I do you'll burst into tears and say that I am just pretending.
You've already forgotten what you said to me this morning. You told me not to
touch you. And so I haven't."

to gender stereotypes and perceived lower negotiation competence, women are more likely to be misled and manipulated by salespeople, irrespective of the salesperson's gender, and by seducers. In one of a series of experiments 24% of men and 11% of women admitted that they had lied to female participants in an exercise of selling real estate. In contrast, only 3% of men and 11% of women lied to men.[12]

On the other hand, men more often interpret spoken language at face value and seem less capable of reading between the lines. Men generally find statements like "*although I said* **ABC**, *the circumstances should have made you understand that I meant* **XYZ**" extremely difficult to comprehend. Men are less receptive to nonverbal messages and rely more on the verbal language than on nonverbal messages. They also seem to implicitly trust verbal language and are less receptive to nonverbal clues.[13]

Men and women not only employ different styles of communication but they place different emphasis on the three often overlapping purposes of a conversation: socialization, exchange of information, and negotiation. Men seem to be especially interested in providing and receiving information. They offer solutions to problems, even if not explicitly asked for them, and are more concerned about status and competition in the framework of a conversation. Women tend to concentrate during a conversation on strengthening a relationship, while focusing on a shared solution to a problem.

If communication tools and content differ to such an extent in men and women, it should not come as a surprise that these same factors apply in conversations between a medical doctor and patient. If men and women speak different "genderlects" it's clear that this would also affect doctor–patient communication. In other words, the gender of both patient and physician and the sex concordance (i.e., doctor and patient are of the same sex) or discordance (different sex) are all likely to influence the interactive process. Indeed this hypothesis is a working assumption that has been substantiated in numerous scientific studies.[14]

HOW DOES THE SEX OF THE DOCTOR AND THE PATIENT AFFECT THE MEDICAL ENCOUNTER?

A mounting body of evidence has shown that better patient-doctor communication improves adherence to treatment recommendations, leads to better disease control and health status, and achieves better patient satisfaction. Yet it's also clear that female and male physicians communicate differently with their patients. Public Health expert Debra Roter and colleagues published a study in which they summarized the results of 26 research projects in which various gender aspects of doctor–patient communication were examined.[15] They focused predominantly on primary care but also on obstetrics and gynecology, pediatric, and internal medicine. They found that female doctors engaged significantly more often in psychosocial discussion with their patients and were more likely to build a partnership with their patient by requesting patient input and assuming a less dominant position within the relationship. Female doctors were more likely to create a positive atmosphere during the visit by verbally agreeing with, encouraging, and reassuring the patient. Female doctors also engaged in more emotionally focused conversation, including asking about feelings and concerns. Finally, female doctors communicated a positive attitude nonverbally through body position (like leaning forward), smiling, and head nods. Male doctors on the other hand were assertive, concentrated on fact gathering, physical examination, and taking the patient's history. They also gave more advice.

One of the immediate practical results of the different attitudes of female and male physicians is that female physicians spend on average 10% more time with their patients. Given the rather restricted time schedule in which all physicians work, it is inevitable that female doctors tend to lag behind in their schedule, and the additional time given to patients during each visit may easily end up as an extra hour or more of work at the end of the day.

GENDER PREFERENCE OF PHYSICIANS

When I entered medical school over 50 years ago, female students were a very small minority of the student body. This gender ratio continued throughout my residency and for quite some time thereafter. Even as a gynecologist and obstetrician for many years, it did not occur to me that some of my patients would have preferred a female doctor. Today, at least in the Western world, there is a clear trend towards a change in the ratio between female and male doctors in favor of women. In a few medical disciplines, such as gynecology, the majority residents are women. Gender preference by patients can inform their choice of doctor, and happily women can choose to see a female Ob-Gyn if they prefer.

Yet in light of the movement toward gender equality among doctors, it's interesting to note that the patient's perception of their doctor continues to be closely tied to the fulfillment of gender norms. As researchers from Switzerland have shown, patients expect different behaviors from female and male doctors.[16] Patients reported experiencing more satisfaction from their doctor's visit if doctors showed stereotypical behavior patterns according to their gender. Specifically, female doctors were expected to show more interest in interpersonal orientation, caring and empathy, be less dominant and not overly assertive while at the same time demonstrating professionalism and signs of status (white coat, stethoscope around the neck, etc.). In an encounter with male physicians, patients expected more distance, more assertiveness, and more adherence to typical male stereotypes. They were less concerned whether male physicians exhibited medical status symbols.

An increasing number of studies indicate that women generally prefer to be seen by a female physician while male patients show no real preference. In a study from the University of California, USA, based on patients who visited an emergency department, 52% of female patients said they trusted their female doctor while only 39% said they trusted their male doctor. Indicators such as "time spent," "concern shown," and "overall care" rated higher for female physicians than for male physicians. On the other hand, male patients did not rate female and male physicians differently in any of the assessed satisfaction indicators.[17] For gynecological exams, female patients, especially in the younger age group (mean age 33.7 years as opposed to 42.8 years),[18] prefer same sex doctors. Men too, prefer male physicians for intimate exams, though they showed a less marked preference. Still, although for some reason most urologists are men, nurses and ultrasound technicians are in most cases women. They most often perform nursing events, technical exams and procedures on men, like catheterization or testicular ultrasound. Men have no choice in this respect, and may suffer embarrassment or even avoid these procedures. Interestingly, children prefer same sex doctors, while their parents are more satisfied when their children are treated by female doctors.

A study conducted in the United States examined the concordance question in 92,000 doctor visits.[19] In disciplines like primary care medicine, psychiatry, dermatology, and pediatrics, female doctors treated between up to 20% more women than men, and this difference was statistically significant. The results also showed that at least in primary care medicine, increasingly women are being treated by female doctors. Numerous studies have shown that female doctors devoted more time to their patients and recommended more preventive medical actions than male physicians did, such as PAP smears for early detection of cervical cancer and mammograms. On the other hand, male physicians invested more time in discussing substance abuse, such as alcohol and drugs.[20]

All of these findings lead to an inevitable question: Do female doctors provide better medical care than male doctors? An extensive study on 160,000 adult diabetes patients treated by family practitioners in the United States was designed to examine the four possible doctor–patient sex pairs and the impact of concordances on control over risk factors.[21] They found that female patients treated by female doctors had better diabetes control than all other doctor–patient dyads (Table 1). Female patients treated by female doctors also exhibited better results for cholesterol level and blood pressure, and were more likely to receive intensive treatment of risk factors than female patients who were treated by male doctors.

On the other hand, another study in the United States that included 5,700 patients being treated for obesity examined the effects of gender concordance on weight-loss recommendations.[22] Male patients treated by male doctors received significantly more advice regarding diet and physical activity than did female patients, and even more than those treated by female doctors.

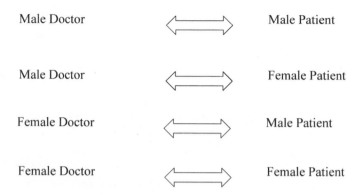

Table 1. doctor–patient sex pairs

Most research shows—as may be expected—that communication in different-sex pairs is more difficult than in same-sex pairs. Researchers from Australia reported that female doctors had greater difficulty relating to male patients than to female patients and that men tended to provide less information than women, particularly regarding their emotional state.[23] In another study gender concordance did not improve health care in patients with depression but counseling had a different focus when provided by female doctors or by male doctors.[24] Female physicians were more likely to counsel on anxiety while male doctors were more likely to counsel on alcohol and drug abuse.

Ultimately, the question of whether female physicians are better than male physicians seems moot. There are areas where female physicians fare better and areas where male physicians fare better. In general, communication seems to be better in women-women pairs which may lead to better treatment results. What's most important is that patients should be given a chance to choose a physician of the same sex if they prefer.

More importantly, the doctor–patient encounter deserves much more attention by the medical system and academia, with a focus on allotting more time to the doctor–patient encounter. The goal should be to further develop Patient Centered Care (PCC) as the desired platform for communication. PCC is a skill and as such, it is teachable. PCC is also measurable. Its tools include videotaping visits and retroactively scoring a variety of interactions between care giver and patient. University curricula and residencies should include the teaching of the required skills to a much greater extent. UC Davis professor and physician Klea Bertakis and colleagues recruited 315 women and 194 men to a study on the effects of patient-centered care.[25] Since female patients are thought to ask more questions and to engage more in conversation, it was assumed that female patients would induce more patient centered care than male patients. This was not the case and the gender of the patients did not affect the score of patient centered care. Obviously, it is the physician's responsibility to develop the framework of PCC.

Changing the basis of the doctor–patient encounter, allocating more time and putting emphasis on a communicative, friendly environment requires appropriate funding and reorganization of resources, including manpower. Since it is beyond the scope of this book to discuss these issues, I would just leave it to medical economics specialists to assess if the required changes might not turn out to be an economically positive cost-benefit investment for the promotion of general public health. It might be that better care can actually be provided at a lower cost to the health care system.[26]

The face of the medical system in the developed world continues to change rapidly. It has become more competitive, more consumer-oriented and at the same time more disease- than patient-oriented. Patients have become more knowledgeable, they are better protected by the legal system, and they have grown increasingly critical. The information revolution has allowed more people to access data, and patients have become more involved in medical decision-making processes. This has certain downsides, as patients can easily become overwhelmed by irrelevant and unsound information.

From the gender perspective, medicine is becoming more feminized and the gender distribution among primary care physicians is rapidly changing. In 2013 there were approximately 770,000 physicians under age 75 in active practice in the USA, about 31% of them female. On the other hand close to 50% of medical students in the years 2010–2011 were women, indicating that in the not-too-far future the ratio between female and male physicians will change in favor of the former.[27] Data from the UK indicate that by 2017 there will be more female physicians than male physicians there.

PCC is an important concept in doctor patient relationships and from the data available it seems that female physicians are somehow better geared than their male counterparts in practicing PCC. It is of great importance that university curricula and residency programs should put much more emphasis on teaching PCC in particular and communication skills in general. This may even be more important as far as male students and residents are concerned.

CHAPTER 15

GENDER MEDICINE—
FUTURE DIRECTIONS

I N THE MID-TWENTIETH CENTURY TWO MAJOR MEDICAL DISAS-ters occurred, both involving treatments for pregnant women and both of which resulted in subsequent catastrophic harm to their children. The first involved a drug called DES (diethyl stilbestrol), a synthetic estrogen developed in the 1930s and administered widely to treat bleeding during pregnancy. Until the 1970s millions of women were treated with DES, even though as early as 1953 a research study indicated that the treatment was ineffective in preventing miscarriages or prema-ture births. By the end of the 1950s, reports of problems in the sexual organs of children borne by women who had been treated with DES during their pregnancies began to appear in the scientific literature. Subsequently, it became clear that DES caused malignancy in the genitalia of offspring and many thou-sands of cases were reported. Finally, in 1971 the FDA issued a bulletin advising physicians to stop prescribing DES to preg-nant women.

The second disaster was related to the drug thalido-mide. Thalidomide was developed initially during the 1950s as a medication to inhibit contractions and treat allergies but it was found to be ineffective for either of these uses. However, it was found to be effective in treating nausea and was pre-scribed to many pregnant women to allay their symptoms. In

1961 two physicians, one from Australia and the other from Germany, reported cases of children born with serious malformations of their limbs, making the connection between these defects and treatment of the mothers with thalidomide. It quickly became clear that thousands of children had been born with similar defects after their mothers had been treated with thalidomide during their pregnancies, and the use of the drug was discontinued. Today thalidomide is used widely and successfully in treating cancer, leprosy, and tuberculosis.

These two disasters shook the world of medicine. As a result, in 1977 the FDA issued a directive that women of childbearing age no longer be included in certain phases of clinical research. This directive was published in order to protect the fetuses from the fate of the DES and thalidomide babies. The new directive was welcomed by researchers and by the pharmaceutical industry, which in any case had preferred to use men in research. Men were more convenient research subjects: they do not have a monthly menstrual period, do not get pregnant, and for the most part are not burdened with taking care of the home, all factors liable to disrupt the course of research. But adoption of this recommendation proved to be much more comprehensive, and in practice even women who were not of childbearing age were excluded from all research. The result was that, in essence, women disappeared from clinical research.

A decade later, the American National Institutes of Health (NIH) attempted to turn back the clock, and in 1986 announced that research that did not include both men and women would no longer be funded. But this statement hardly changed the practice of using men for research. In 1990 the NIH established a department intended to promote research on women and even funded the largest research study ever conducted in this field, which included around 160,000 postmenopausal women at a cost of more than 620 million dollars.[1]

In 1993 the NIH directive received legal backing, and the funding of research that did not include women was declared as unacceptable by all American government institutions. Today, in research studies funded by public funds in the United States the number of men and women is practically equal. Yet while the law obliges researchers to include women in their studies, in practice they neglect to analyze the results in accordance with the participants' sex. Thus this law is of little practical value. A 2004 article surveyed studies published in nine leading scientific journals. The survey examined how many of those research studies that received funding from the US government analyzed the results according to sex and gender, and if they did not, whether a sufficient explanation was given for failing to do so. No such analysis was performed in 67% of the articles surveyed, and no explanation was given why not. In 2011, the same research group published a followup study that examined the same questions in the same journals. This follow-up study again revealed that 64% of the studies published in the leading medical journals disregarded the requirement to include women in clinical research or to analyze the results according to sex and gender.[2] Furthermore, even in animal studies researchers prefer to use male animals, among other reasons because of the females' hormonal cycles. Researchers in the United States examined around two thousand medical studies on animals published in 2009 and found that in eight out of ten fields of medicine more males were included in the experiments than females. Even in studies of diseases whose incidence is much higher in women, the animal experiments used primarily male animals.[3]

• • •

Why is it so difficult to bring about change? Despite the tremendous progress in the field of medicine today, many doctors find it difficult to accept changes in their everyday practice unless they had hoped for this particular change themselves. Two examples from obstetrics illustrate this point.

One of the most serious problems related to prematurely born babies is that their lungs are not fully developed. In many cases this leads to serious illness and mortality. Between 1972 and 1982 seven research studies were published that noted the benefits of administering steroid drugs to the mother in order to help her baby's lungs develop and mature. Within the next decade, an additional seven studies were published indicating a decrease in mortality of 30–50% among premature babies when mothers followed this treatment. Even though these research studies were published in excellent scientific journals, it was only in an NIH conference in 1994 that experts published their recommendation regarding this treatment. Following this publication, use of this treatment method rose from 21% to 46%. However, that meant that more than half of obstetricians were still not using this lifesaving treatment. Apparently, the availability of novel and scientifically sound information on a new treatment is not always a sufficient incentive to apply such knowledge. More data does not necessarily help change habits.

My second example shows the opposite tendency and involves the rejection of vaginal deliveries when the baby is in breech presentation—that is, when the baby's buttocks face the birth canal—in favor of Caesarean sections. A large research study published in 2000 apparently demonstrated that for babies with breech presentations a Caesarean section may be safer than a vaginal delivery.[4] International professional organizations quickly adopted this conclusion and issued recommendations which led to a mass abandonment almost overnight of planned vaginal breech births in the Western world. The reasons for this almost uniform acceptance by obstetricians were related not only to the recommendations of the professional organizations but also to the fact that many obstetricians already preferred birth by Caesarean section. Greater expertise is required for vaginal breech deliveries than for Caesarean sections; Caesarean sections can be scheduled in advance; and more lawsuits are filed against doctors for complications of a vaginal birth than for complications of a Caesarean section. The doctor's compensation for performing a Caesarean section is also greater than for a vaginal birth, at least in the United States. (Nevermind the longer recovery time for mothers who undergo a Caesarian section, risks involved in surgery and rising risks in subsequent pregnancies. Thus it may be that the medical establishment was simply waiting for scientific confirmation in order to abandon vaginal births with breech presentation.

• • •

Several years later it emerged that the methodology used in the research that began this avalanche was fraught with serious failings. Many published opinions and research studies, including a reanalysis of the data on which the original research had been based, refuted the original conclusions.[5] But it was too late. A new reality had already been established. After six years during which only isolated centers and obstetricians still practiced planned vaginal births for breech presentation, virtually no residents were taught the skills required for this procedure, and in view of obstetricians' underlying lack of desire to perform such births, there was no way back. Today most hospitals in the Western world lack sufficient expertise to perform vaginal breech births. Hence, the number of unnecessary

Caesarean sections has risen, and along with it the rate of complications. We find ourselves in an untenable position: birth by Caesarean section is not beneficial for a fetus in breech presentation and it significantly increases the risk to the mother, particularly for subsequent births. Yet for the most part obstetricians no longer have the skills or motivation necessary for vaginal delivery with breech presentation.

As such examples show, changes in medicine do not necessarily occur based upon the presence of solid scientific evidence and may even occur in spite of robust evidence. In order to bring about change we must scrupulously consider the appropriate tools.

In order for gender medicine to be introduced and implemented at clinics and hospitals, we must first and foremost raise awareness. Heightened awareness may lead to policy changes. Beyond that there is a need for conceptual changes that require the willingness to change paradigms and redirect investment in research and teaching. Most physicians are not indifferent when it comes to gender medicine, as they were in the case of the treatment for lung maturation in premature fetuses. It is rare to find a professional who, after learning about the principles of gender medicine, still rejects or disregards the idea. But to get from hazy awareness to concrete implementation we still have a long way to go.

WHERE DO WE STAND TODAY?

Today Gender and Sex Specific Medicine is accepted as an undeniable and necessary paradigm shift, not least to reverse the harms of the 1977 FDA directive. This shift was put into motion quite some time ago. The Office on Women's Health of the U.S. Department of Health and Human Services was established in

1990 and has invested tremendous funds and efforts to promote this field. The European commission is also funding research programs and scientific meetings as are many European governmental agencies. National and international scientific meetings are being held worldwide. As of this writing, eight national societies devoted to Gender and Sex Specific Medicine (Austria, Germany, Israel, Italy, Japan, Sweden, USA) have joined in the International Society for Gender Medicine, which acts as an umbrella organization with over 720 members (*www.isogem.com*).

Many scientific organizations are recognizing the value of gender-specific medical research as well. NASA has done pioneering work in studying the physiological differences between men and women as related to space travel. An increasing number of medical schools have integrated the study of Gender and Sex Specific Medicine (GSSM) into their core curriculum. Fellowships in women's health are offered by various universities. Publications of basic and clinical research are mushrooming and research centers dedicated to GSSM are being established in increasing numbers at leading academic institutions worldwide. Books have been written on the topic for the general public and a number of textbooks on GSSM are available as well.[6,7,8,9] There are scientific journals devoted to GSSM and leading scientific journals have made it their policy to publish research only if data are stratified by gender or if a satisfactory explanation is provided when stratification has not been made.

RESEARCH IN GENDER MEDICINE

Research in gender medicine aims to investigate the functioning of the bodily systems of men and women and the distinctions between them so as to gain a deeper understanding of the topics we've covered in this book. It aims to examine why certain diseases have different manifestations in the two sexes, to study the reasons for the high incidence of certain diseases in one of the sexes, to examine distinctive functional mechanisms of bodily systems in the two sexes and to study the differential influences of the same medication on the two sexes. In addition, research in gender medicine must place emphasis on the difference between the sexes in animal experimentation and in cell research.

Gender medicine, then, is the logical development of the aspiration to improve the quality of evidence-based medicine. The objective of evidence-based medicine is to use precise analysis to identify the factual strength of research studies in order to assess their results. This involves examining the structure of the research, the method used to gather the data and the statistical methods used to examine the results. For this reason, studies seeking to prove the diagnostic or therapeutic benefits of a particular treatment that do not include analysis of the results for each of the two sexes separately cannot be considered evidence-based. There is no doubt in my mind that a better understanding of the biological differences between the sexes will lead to an improvement of the medical services given to both women and men.

HOW CAN WE PROMOTE GENDER MEDICINE?

Gender medicine is still a young field and most physicians are not familiar with its principles. Thus we need to start at the grassroots and take gender medicine's principles into the doctor's office. While asking our doctor to take gender aspects into consideration may cause him or her discomfort and stress, on the other hand, the question may motivate the physician to learn more about this topic. And the more doctors learn about gender medicine, the more doctors will be equipped to answer such questions.

Changing reality to the point of developing a grassroots movement is highly dependent on the education and awareness of the general public and the media about gender medicine, but I believe we can make it happen. Patients need to address physicians and medical institutions need to pay attention to gender issues. Regulators such as health ministries, legislative bodies, and public health institutions will need to understand the extent of their responsibility in promoting gender medicine, in encouraging its development, in advancing research and teaching about it and in allocating resources for these purposes. The pharmaceutical industry will need to invest in research related to medications and prepare for the time—I hope not in the distant future—when pharmacy shelves will be stocked with different drug preparations for men and women.

Internationally, more scientific organizations will need to be founded in order to further promote research, teaching, and implementation of GSSM. Will GSSM become a separate medical discipline such as pediatrics or gynecology? Probably not. There will probably be a GSSM super-discipline that will deal with interdisciplinary gender issues but GSSM should be part of all medical disciplines in teaching research and medical services.

In the not-so-distant future I anticipate that gender medicine will completely transform medical care so that each of us receives the individual, evidence-based treatment we deserve.

ACKNOWLEDGMENTS

The story of this book begins with a lecture series, which I gave on Israeli radio for the University of Tel Aviv in 2013 and 2014. I owe thanks to Dr. Boaz Hagay, the gifted producer and editor of this venerable institution, who pushed me to expand, update, and develop the series into a book. The book has become quite a success in Israel, and patients often give it to their doctors. Thanks to these patients for stimulating their physicians to learn more about this exciting new science.

Thanks are due to Donna Bossin for translating the Hebrew text into English. I am deeply indebted to my friend Deborah Owen, who helped me tremendously with her wise advice and made contact with Peter Mayer at The Overlook Press. Peter and Tracy Carns did a fabulous editing job. Big thanks to both of them. I'm also very grateful to my dear friend, the distinguished Israeli writer Amos Oz, for his critical comments and guidance, for allowing me to use citations from one of his novels, and for writing the beautiful introduction to this book.

Thanks to Dr. Marianne Legato, a pioneer in the field of gender medicine, whose books were eye-openers to me. Thanks to Doctor Eran Halpern, CEO of Rabin Medical Center, whose continuous help and support have allowed me to promote gender medicine in Israel. Thanks to the former dean of the Sackler School of Medicine at Tel Aviv University, Professor Yoseph Mekori, and to the current dean, Professor Ehud Grossman for their vision. Thanks to genetician Professor Motti Shohat, cardiologist; Doctor Avital Porter; neurologist Doctor Arie Kuritzki; and gastroenterologist Professor Ram Dickman for their expertise. Thanks to Ilit Shefer for her wonderful and clear illustrations.

And on a very personal note: Before I retired from my clinical duties in 2012, I had promised to devote much more time to my family, to whom this book is dedicated. I am very grateful to them for forgiving me for postponing that promise and allowing me to work on *Gender Medicine*. Writing this book was a fabulous experience. I have been a gynecologist and obstetrician for my entire professional life, but very little of this book deals with my specialty, so I had to do background research in many different medical areas. I dove deep into various medical specialties and learned about new developments and discoveries. My greatest and most rewarding personal experience throughout this voyage was to enjoy the excitement of curiosity. So, my deepest, humble thanks to destiny, which sent me on this journey of exploring gender medicine.

ENDNOTES

Introduction

1. Kurzweil, R. (2005). *The Singularity is Near.* New York, NY: Penguin.
2. Craft, R. M., et al. (2004). Sex differences in pain and analgesia: The role of gonadal hormones. *European Journal of Pain, 8,* 397–411.
3. Shaywitz, B. A., et al. (1995). Sex differences in the functional organization of the brain for language. *Nature, 373,* 607–609.
4. Kent, D. M., et al. (2005). Sex-based differences in response to recombinant tissue plasminogen activator in acute ischemic stroke: A pooled analysis of randomized clinical trials. *Stroke, 36,* 62–65.
5. Bornstein, M. H., et al. (2000). Child language with mother and with stranger at home and in the laboratory: A methodological study. *Journal of Child Language, 27,* 407–442.
6. Murray, A. D., et al. (1990). Fine-tuning of utterance length to preverbal infants: Effects on later language development. *Journal of Child Language, 17,* 511–525.
7. Roulstone, S., et al. (2003). The speech and language of children aged 25 months: Descriptive data from the Avon longitudinal study of parents and children. *Early Child Development & Care, 172,* 259–268.
8. Voskuhl, R. (2011). Sex differences in autoimmune diseases. *Biology of Sex Differences, 2.* doi: 10.1186/2042-6410-2-1
9. Yoshida, R., et al. (2003). Motion and morphology of the thumb metacarpophalangeal joint. *Journal of Hand Surgery, 28,* 753–757.

10. Toth, A. P., et al. (2001). Anterior cruciate ligament injuries in the female athlete. *Journal of Gender Specific Medicine, 4,* 25–34.

Chapter 1

1. Courtenay, W. (2000). Constructions of masculinity and their influence on men's well-being: A theory of gender and health. *Social Science & Medicine, 50,* 1385–1401.

2, Warner, D. A., & Shine, R. (2009). The adaptive significance of temperature-dependent sex determination in a reptile. *Nature, 451,* 566–568.

3. Warner, R. R., & Swearer, S. E. (1991). Social control of sex change in the bluehead wrasse, Thalassoma bifasciatum (Pisces: Labridae). *Biological Bulletin, 181,* 199–204.

4. Balance Systems, Inc. (N.d.). National and international statistics for carpal tunnel syndrome and repetitive strain injuries of the upper extremity. Retrieved from http://www.repetitive-strain.com/national.html

5. Van Rijn, R. M., et al. (2009). Associations between work-related factors and the carpal tunnel syndrome—A systematic review. *Scandinavian Journal of Work, Environment & Health, 35,* 19–36.

6. Fan, F. C., et al. (2012). Non-invasive prenatal measurement of the fetal genome. *Nature, 487,* 320–324.

7. Mungall, A. J. (2002). Meeting review: Epigenetics in development and disease. *Comparative & Functional Genomics, 3,* 277–281.

8. Ngalamika, O., et al. (2012). Epigenetics, autoimmunity and hematologic malignancies: A comprehensive review. *Journal of Autoimmunity, 39,* 451–465.

9. Yang, X., et al. (2006). Tissue-specific expression and regulation of sexually dismorphic genes in mice. *Genome Research, 16,* 995–1004.

10. McGowen, M. L. (2014). Integrating genomics into clinical oncology: Ethical and social challenges from proponents of personalized medicine. *Urologic Oncology: Seminars & Original Investigations, 32,* 187–192.

11. Li, C. (2011). Personalized medicine—The promised land: Are we there yet? *Clinical Genetics, 79,* 403–412.

Chapter 2

1. Donald, I. (1969). Sonar as a Method of Studying Prenatal Development. *Journal of Pediatrics, 75,* 326–333.

2. Huxley, A. (1932). *Brave New World.* New York, NY: Doubleday, Doran.

3. Lesseur, C., et al. (2015). Sex-specific associations between placental leptin promoter DNA methylation and infant neurobehavior. *Psychoneuroendocrinology, 40,* 1–9.

4. Gluckman, P., et al. (2005). *The Fetal Matrix: Evolution, Development and Disease.* New York, NY: Cambridge University Press.

5. Hiramatsu, A. (2009). Critical time window of SRY action in gonadal sex determination in mice. *Development, 136,* 129–138.

6. Goodfellow, P. N., et al. (1993). SRY and sex determination in mammals. *Annual Review of Genetics, 27,* 71–92.

7. Veitia, R. A. (2010). FOXL2 versus SOX9: A Lifelong "Battle of the Sexes." *Bioessays, 32,* 375–380.

8. Agate, R. J., et al. (2003). Neural, not gonadal, origin of brain sex differences in a gynandromorphic finch. *Proceedings of the National Academy of Sciences, 100,* 4873–4878.

9. Ibid.

10. Ben-Haroush, A., et al. (2012). Early first-trimester crown-rump length measurements in male and female singleton fetuses in IVF pregnancies. *Journal of Maternal-Fetal & Neonatal Medicine, 25,* 2610–2612.

11. Phoenix, C. H., et al. (1959). Organizing action of prenatally administered testosterone proprionate on the tissues mediating mating behavior in the female guinea pig. *Endocrinology, 65,* 369–382.

12. Nordenström, A., et al. (2002). Sex-typed toy play behavior correlates with the degree of prenatal androgen exposure assessed by CYP21 genotype in girls with congenital adrenal hyperplasia. *Journal of Clinical Endocrinology & Metabolism, 87,* 5119–5124.

13. Baron-Cohen, S., et al. (2004). *Prenatal Testosterone in Mind: Amniotic Fluid Studies.* Cambridge, MA: MIT Press.

14. Sapienza, P., et al. (2009). Gender differences in financial risk aversion and career choices are affected by testosterone. *Proceedings of the National Academy of Sciences, 106,* 15268–15273.

15. Chapman, E., et al. (2006). Fetal testosterone and empathy: Evidence from the empathy quotient (EQ) and the "Reading the Mind in the Eyes" test. *Social Neuroscience, 1,* 135–148.

16. Desai, M., et al. (2005). Programmed obesity in intrauterine growth-restricted newborns: Modulation by newborn nutrition. *American Journal of Physiology, 288,* 91–96.

17. Lee, T. M., et al. (1988). Vole infant development is influenced perinatally by maternal photoperiodic history. *American Journal of Physiology, 255,* 831–838.

18. Mennella, J. A., et al. (2001). Prenatal and postnatal flavor learning by human infants. *Pediatrics, 107,* 1–6.

19. Trivers, E., & Willard, D. E. (1973). Natural selection of parental ability to vary the sex ratio of offspring. *Science, 179,* 90–92.

20. Obel, C., et al. (2007). Psychological distress during early gestation and offspring sex ratio. *Human Reproduction, 22,* 3009–3012.

21. Cameron, E. Z., et al. (2009). A Trivers-Willard effect in con-

temporary humans: Male-biased sex ratios among billionaires. *PLOS ONE, 4.* doi: 10.1371/journal.pone.0004195

22. Catalano, R. A., et al. (2006). Secondary sex ratios and male lifespan: Damaged or culled cohorts. *Proceedings of the National Academy of Sciences, 103,* 1639–1643.

23. Devlin, B., et al. (1997). The heritability of IQ. *Nature, 388,* 468–471.

24. Turkheimer, E. (2003). Socioeconomic status modifies heritability of IQ in young children. *Psychological Science, 14,* 623–628.

25. Chura, L. R., et al. (2010). Organizational effects of fetal testosterone on human corpus callosum size and asymmetry. *Psychoneuroendocrinology, 35,* 122–132.

26. Gould, E., et al. (1999). Neurogenesis in the neocortex of adult primates. *Science. 286,* 548–552.

27. Agin, D. (2010). *More than Genes.* Oxford, England: Oxford University Press.

28. Clements, M. A., et al. (2006). Sex differences in cerebral laterality of language and visuospatial processing. *Brain and Language, 98,* 150-8.

29. Bowers, J. M., et al. (2013). Foxp2 mediates sex differences in ultrasonic vocalization by rat pups and directs order of maternal retrieval. *Journal of Neuroscience, 33,* 3276–3283.

30. Ingalhalikar, M., et al. (2014). Sex differences in the structural connectome of the human brain. *Proceedings of the National Academy of Sciences, 111,* 823–828.

31. Joel, D., et al. (2015). Sex beyond the genitalia: The human brain mosaic. *Proceedings of the National Academy of Sciences, 112,* 15468–15473.

32. Glezerman, M. (2016). Yes, there is a female and a male brain: Morphology versus functionality. *Proceedings of the National Academy of Sciences USA.* 2016 Mar 8. pii: 201524418. [Epub ahead of print] PMID: 26957594

Chapter 3

1. Nugent, B. M., et al. (2015). The omniscient placenta: Metabolic and epigenetic regulation of fetal programming. *Frontiers in Neuroendocrinology, 39,* 28–37.

2. Whyatt, R. M., et al. (2002). Residential pesticide use during pregnancy among a cohort of urban minority women. *Environmental Health Perspectives, 110,* 507–514.

3. Anway, M. D., et al. (2006). Epigenetic transgenerational actions of endocrine disruptors. *Endocrinology, 147,* S43–S49.

4. Agin, D. (2010). More than genes. Oxford, England: Oxford University Press.

5. Ibid.

6. U.S. Department of Health and Human Services. (2005, February 21). U.S. surgeon general releases advisory on alcohol use in pregnancy. Retrieved from http://come-over.to/FAS/Sur GenAdvisory.htm

7. Sittig, L. J., et al. (2011). Strain-specific vulnerability to alcohol exposure in utero via hippocampal parent-of-origin expression of deiodinase–III. *Federation of American Societies for Experimental Biology Journal, 25,* 2313–2324.

8. Xiao, D., et al. (2008). Prenatal gender-related nicotine exposure increases blood pressure response to angiotensin II in adult offspring. *Hypertension, 51,* 1239–1247.

9. Toro, R., et al. (2008). Prenatal exposure to maternal cigarette smoking and the adolescent cerebral cortex. *Neuropsychopharmacology, 33,* 1019–1027.

10. Jacobsen, L. K., et al. (2007). Gender-specific effects of prenatal and adolescent exposure to tobacco smoke on auditory and visual attention. *Neuropsychopharmacology, 32,* 2453–2464.

11. Hines, M., et al. (2004). Androgen and psychosexual development: Core gender identity, sexual orientation and recalled childhood gender role behavior in women and men with con-

genital adrenal hyperplasia (CAH). *Journal of Sex Research, 41*, 1–7.

12. Hu, M., et al. (2015). Maternal testosterone exposure increases anxiety-like behavior and impacts the limbic system in the offspring. *Proceedings of the National Academy of Sciences, 112*, 14348–14353.

13. Baron-Cohen, S., et al. (1997). Is autism an extreme form of the "male brain"? *Advances in Infancy Research, 11*, 193–217.

14. Devlin B., et al. (1997). The heritability of IQ. *Nature, 388*, 468–471.

15. Rabinowitz, M. (2009). *Obesity in Israel*. Report to the Israel Knesset.

16. World Health Organization and UNICEF. (2004). *Low Birthweight: Country, Regional and Global Estimates*. Retrieved from www.childinfo.org/files/low_birthweight_from _EY.pdf

17. Barker, D. J., et al. (1993). Fetal nutrition and cardiovascular disease in adult life. *Lancet, 341*, 938–941.

18. Sardinha, F. L. C., et al. (2006). Gender difference in the effect of intrauterine malnutrition on the central anorexigenic action of insulin in adult rats. *Nutrition, 22*, 1152–1161.

19. Tamimi, R. M., et al. (2003). Average energy intake among pregnant women carrying a boy compared with a girl. *BMJ, 326*, 1245–1246.

20. Leeson, C. P. M., et al. (2001). Impact of low birth weight and cardiovascular risk factors on endothelial function in early adult life. *Circulation, 103*, 1264–1268.

21. Lussana, F., et al. (2008). Prenatal exposure to the Dutch famine is associated with a preference for fatty foods and a more atherogenic lipid profile. *American Journal of Clinical Nutrition, 88*, 1648–1652.

22. Neel, J. V. (1962). Diabetes mellitus: A "thrifty" genotype ren-

dered detrimental by "progress"? *American Journal of Human Genetics, 14,* 353–362.

23. Hales, C. N., & Barker, D. J. (1992). Type 2 (non-insulin-dependent) diabetes mellitus: The thrifty phenotype hypothesis. *Diabetologia, 35,* 595–601.

24. Stride, A., & Hattersley, A. T. (2002). Different genes, different diabetes: Lessons from mature onset diabetes of the young. *Annals of Medicine, 34,* 207–216.

25. Eriksson, J. G., et al. (2002). The effects of the Prof12Ala polymorphism of the peroxisome proliferator-activated receptor-y2 gene on insulin sensitivity and insulin metabolism interact with size at birth. *Diabetes, 51,* 2321–2324.

26. Kaseva, N., et al. (2012). Lower conditioning leisure-time physical activity in young adults born preterm at very low birth weight. *PLOS ONE, 7.* doi: 10.1371/journal.pone.0032430

27. Ross, M. G., et al. (2013). Developmental programming of offspring obesity, adipogenesis, and appetite. *Clinical Obstetrics & Gynecology, 56,* 529–536.

28. Portella, A. K., et al. (2012). Effects of in utero conditions on adult feeding preferences. *Journal of Developmental Origins of Health & Disease, 3,* 140–152.

29. Barbieri, M. A., et al. (2009). Severe intrauterine growth restriction is associated with higher spontaneous carbohydrate intake in young women. *Pediatric Research, 65,* 215–220.

30. Hales, C. N., et al. (2003). The dangerous road of catch-up growth. *Journal of Physiology, 547,* 5–10.

31. Nathanielsz, P. (1999). *Life in the Womb: The Origin of Health and Disease.* Ithaca, NY: Promethean.

Chapter 4

1. Paul, A. M. (2010). *Origins: How the First Nine Months before Our Birth Shape the Rest of Our Lives.* New York, NY: Free Press.

2. Monk, C., et al. (2003). Effects of women's stress-elicited phys-iological activity and chronic anxiety on fetal heart rate. *Journal of Developmental & Behavioral Pediatrics, 24*, 32–38.

3. Ward, I. L. (1972). Prenatal stress feminizes and demasculinizes the behavior of males. *Science, 7*, 82–84.

4. Van den Bergh, B. R., et al. (2004). High antenatal maternal anxiety is related to ADHD symptoms, externalizing problems, and anxiety in 8- and 9-year-olds. *Child Development, 75*, 1085–1097.

5. Bergman, K., et al. (2007). Maternal stress during pregnancy predicts cognitive ability and fearfulness in infancy. *Journal of the American Academy of Child & Adolescent Psychiatry, 46*, 1454–1463.

6. DiPietro, J. A., et al. (2006). Maternal psychological distress during pregnancy in relation to child development at age two. *Child Development, 77*, 573–587.

7. Radtke, K. M., et al. (2011). Transgenerational impact of inti-mate partner violence on methylation in the promoter of the glucocorticoid receptor. *Translational Psychiatry, e2*, 1–6.

8. Khashan, A. S., et al. (2012). Prenatal stress and risk of asthma hospitalization in the offspring: A Swedish population-based study. *Psychosomatic Medicine, 74*, 635–641.

9. Khashan, A. S., et al. (2008). Higher risk of offspring schizo-phrenia following antenatal maternal exposure to severe ad-verse life events. *Archives of General Psychiatry, 65*, 146–152.

10. Hansen D., et al. (2000). Serious life events and congenital mal-formations: A national study with complete follow-up. *Lancet, 356*, 875–880.

11. Van Os, J., et al. (1988). Prenatal exposure to maternal stress and subsequent schizophrenia: The May 1940 invasion of the Netherlands. *British Journal of Psychiatry, 172*, 324–326.

12. Malaspina, D., et al. (2008). Acute maternal stress in pregnancy

and schizophrenia in offspring: A cohort prospective study. *BMC Psychiatry 8.* doi: 10.1186/1471-244X-8-71

13. King, S., et al. (N.d.). Project Ice Storm. Retrieved from https://www.mcgill.ca/projetverglas/icestorm

14. Laplante, P., et al. (2008). Project Ice Storm: Prenatal maternal stress affects cognitive and linguistic functioning in 5 1/2-year-old children. *Journal of the American Academy of Child & Adolescent Psychiatry, 47,* 1063–1072.

15. Weinstock, M. (2007). Gender differences in the effects of prenatal stress on brain development and behaviour. *Neurochemical Research, 32,* 1730–1740.

16. Austin, M. P., et al. (2005). Prenatal stress, the hypothalamic-pituitary-adrenal axis, and fetal and infant neurobehaviour. *Early Human Development, 81,* 917–926.

17. Kleinhaus, K., et al. (2013). Prenatal stress and affective disorders in a population birth cohort. *Bipolar Disorders, 15,* 92–99.

18. Benediktsson, R., et al. (1997). Placental 11 beta-hydroxysteroid dehydrogenase: A key regulator of fetal glucocorticoid exposure. *Clinical Endocrinology, 46,* 161–166.

19. Holmes, M. C., et al. (2006). The mother or the fetus? 11Beta-hydroxysteroid dehydrogenase type 2 null mice provide evidence for direct fetal programming of behavior by endogenous glucocorticoids. *Journal of Neuroscience, 26,* 3840–3844.

20. McCalla, C. O., et al. (1998). Placental 11[beta]-hydroxysteroid dehydrogenase activity in normotensive and pre-eclamptic pregnancies. *Steroids, 63,* 511–515.

21. Van den Bergh, B. R. G., et al. (2004). High antenatal maternal anxiety is related to ADHD symptoms, externalizing problems, and anxiety in 8- and 9-year olds. *Child Development, 75,* 1085–1097.

22. Rice, F., et al. (2010). The links between prenatal stress and off-

spring development and psychopathology: Disentangling environmental and inherited influences. *Psychological Medicine, 40,* 335–345.

23. Hanley, G. E., et al. (2014). The effect of perinatal exposures on the infant: Antidepressants and depression. *Best Practice & Research Clinical Obstetrics & Gynaecology, 28,* 37–48.

24. Bonnin, A., et al. (2011). A transient placental source of serotonin for the fetal forebrain. *Nature, 472,* 347–350.

25. Entringer, S., et al. (2013). Maternal psychosocial stress during pregnancy is associated with newborn leukocyte telomere length. *American Journal of Obstetrics & Gynecology, 208,* 134.e1–134.e7.

26. Armanios, M., al. (2012). The telomere syndromes. *Nature Reviews Genetics, 13,* 693–704.

Chapter 5

1. Avraham, R. (2000). *The Circulatory System.* Philadelphia, PA: Chelsea House.

2. Lampe, F.C., et al. (2000). The natural history of prevalent ischaemic heart disease in middle-aged men. *European Heart Journal, 21,* 1052–1062.

3. Rivera, C. M., et al. (2009). Increased cardiovascular mortality after early bilateral oophorectomy. *Menopause, 16,* 15–23.

4. Parker, W. H., et al. (2009). Ovarian conservation at the time of hysterectomy and long-term health outcomes in the nurses' health study. *Obstetrics & Gynecology, 113,* 1027–1037.

5. Manson, J. E., et al. (2003). Estrogen plus progestin and the risk of coronary heart disease. *New England Journal of Medicine, 349,* 523–534.

6. Maas, A. H., et al. (2011). Red alert for women's heart: The urgent need for more research and knowledge on cardiovascular disease in women. *European Heart Journal, 32,* 1362–1368.

7. Stramba-Badiale, M. (2010). Women and research on cardio-vascular diseases in Europe: A report from the European Heart Health Strategy (EuroHeart) project. *European Heart Journal, 31*, 1677–1681.

8. Grundtvig, M., et al. (2009). Sex-based differences in premature first myocardial infarction caused by smoking: Twice as many years lost by women as by men. *European Journal of Cardiovascular Prevention & Rehabilitation, 16*, 174–179.

9. Huxley, R., et al. (2006). Excess risk of fatal coronary heart disease associated with diabetes in men and women: Meta-analysis of 37 prospective cohort studies. *BMJ, 332*, 73–78.

10. Maas, A.H., et al. (2009). Women's health in menopause with a focus on hypertension. *Netherlands Heart Journal, 17*, 69–73.

11. Lorell, B. H., et al. (2000). Left ventricular hypertrophy: Pathogenesis, detection, and prognosis. *Circulation, 102*, 470–479.

12. Pope, J. H., et al. (2000). Missed diagnoses of acute cardiac ischemia in the emergency department. *New England Journal of Medicine, 342*, 1163–1170.

13. Dey, S., et al. (2009). Sex-related differences in the presentation, treatment and outcomes among patients with acute coronary syndromes: The Global Registry of Acute Coronary Events. *Heart, 95*, 20–26.

14. Tobin, J. N., et al. (1987). Sex bias in considering coronary bypass surgery. *Annals of Internal Medicine, 107*, 19–25.

15. Crilly, M., et al. (2007). Gender differences in the clinical management of patients with angina pectoris: A cross-sectional survey in primary care. *BMC Health Services Research, 7*, 142.

16. Isorni, M. A., et al. (2015). Impact of gender on use of revascularization in acute coronary syndromes: the national observational study of diagnostic and interventional cardiac

catheterization (ONACI). *Catheterization & Cardiovascular Interventions, 86,* E58–E65.

17. McMurray, J. J. V., et al. (2012). ESC guidelines for the diagnosis and treatment of acute and chronic heart failure. *European Heart Journal, 33,* 1787–1847.

18. Klempfner, R., et al. (2014). The Israel Nationwide Heart Failure Survey: Sex differences in early and late mortality for hospitalized heart failure patients. *Journal of Cardiac Failure, 20,* 193–198.

19. Bougouin, W., et al. (2015). Gender and survival after sudden cardiac arrest: A systematic review and meta-analysis. *Resuscitation, 94,* 55–60.

20. Regitz-Zagrosek, V. (2006). Therapeutic implications of the gender-specific aspects of cardiovascular disease. *Nature Reviews Drug Discovery, 5,* 425–438.

21. Murakami, T., et al. (2015). Gender differences in patients with takotsubo cardiomyopathy: Multi-center registry from Tokyo CCU Network. *PLOS ONE, 28.* doi: 10.1371/journal.pone .0136655

22. Retnakaran, R., et al. (2010). Glucose intolerance in pregnancy and postpartum risk of metabolic syndrome in young women. *Journal of Clinical Endocrinolology & Metabolism, 95,* 670–677.

23. Kessous, R., et al. (2013). An association between gestational diabetes mellitus and long-term maternal cardiovascular morbidity. *Heart, 99,* 1118–1121.

24. Kessous, R., et al. (2013). An association between preterm delivery and long-term maternal cardiovascular morbidity. *American Journal of Obstetrics & Gynecology, 368,* e1–e8.

25. Shalom, G., et al. (2013). Is preeclampsia a significant risk factor for long-term hospitalizations and morbidity? *Journal of Maternal-Fetal & Neonatal Medicine, 26,* 13–5.

Chapter 6

1. Chang, L., et al. (2006). Gender, age, society, culture and the patient's perspective in the functional gastrointestinal disorders. *Gastroenterology, 130*, 1435–1446.

2. Camilleri, M., et al. (1999). Improvement in pain and bowel function in in female irritable bowel patients with alosetron, a 5-HT3 receptor antagonist. *Alimentary Pharmacology & Therapeutics, 13*, 1149–1159.

3. Ibid.

4. Sperber, A. D., et al. (2005). Rates of functional bowel disorders among Israeli bedouins in rural areas compared to those who moved to permanent towns. *Clinical Gastroenterology & Hepatology, 3*, 342–348.

5. Wang, C., et al. (2014). Effect of drinking on all-cause mortality in women compared with men. *Journal of Women's Health, 23*, 373–381.

6. Dickman, R., et al. (2014). Gender aspects suggestive of gastroparesis in patients with diabetes mellitus: A cross-sectional survey. *BMC Gastroenterology, 14*. doi: 10.1186/1471-230X-14-34

Chapter 7

1. Gershon, M. D. (1998). *The Second Brain*. New York, NY: Harper.

2. Bayliss, W. M., & Starling, E. H. (1899). The movements and innervation of the small intestine. *Journal of Physiology, 24*, 99–143.

3. Yano, J. M., et al. (2015). Indigenous bacteria from the gut microbiota regulate host serotonin biosynthesis. *Cell, 161*, 264–276.

4. Shannon, K. M., et al. (2012). Alpha-synuclein in colonic submucosa in early untreated Parkinson's disease. *Movement Disorders, 27*, 709–715.

5. Milo, R., et al. (2015). *Cell biology by the numbers*. New York, NY: Garland Science.

6. Pflughoeft, K. J., et al. (2012). Human microbiome in health and disease. *Annual Review of Pathology, 7*, 99–122.

7. Ardissone, A. N., et al. (2014). Meconium microbiome analysis identifies bacteria correlated with premature birth. *PLOS ONE, 9*. doi: 10.1371/journal.pone.0090784

8. Zilber-Rosenberg, I., et al. (2008). Role of microorganisms in the evolution of animals and plants: The hologenome theory of evolution. *FEMS Microbiology Reviews, 32*, 723–735.

9. Blaser, M. (2013). The microbiome explored: Recent insights and future challenges. *Nature Reviews Microbiology, 11*, 213–217.

10. Suez, S. I., et al. (2014). Artificial sweeteners induce glucose intolerance by altering the gut microbiota. *Nature, 514*, 181–186.

11. Trompette, A., et al. (2014). The microbiota metabolism of dietary fiber influences allergic airway disease and hematopoiesis. *Nature Medicine, 20*, 159–166.

12. Ibid.

13. Bolnick, D. I. (2014). Individual diet has sex-dependent effects on vertebrate gut microbiota. *Nature Communications, 5*, 1–12.

14. Biedermann, L. (2013). Smoking cessation induces profound changes in the composition of the intestinal microbiota in humans. *PLOS ONE, 8*. doi: 10.1371/journal.pone.0059260

15. Belkaid, Z., et al. (2014). Role of the microbiota in immunity and inflammation. *Cell, 157*, 121–141.

16. Holbreich, M., et al. (2012). Amish children living in northern Indiana have a very low prevalence of allergic sensitization. *Journal of Allergy & Clinical Immunology, 129*, 1671–1673.

17. Markle, J. G. M., et al. (2009). Sex differences in the gut microbiome drive hormone-dependent regulation of autoimmunity. *Science, 339*, 1084–1088.

18. Zhang, F., et al. (2012). Should we standardize the 1,700 year

old fecal microbiota transplantation? *American Journal of Gastroenterology, 107,* 1755.

19. Van Nood, E., et al. (2013). Duodenal infusion of donor feces for recurrent clostridium difficile. *New England Journal of Medicine, 368,* 407–415.

20. Ding, T., et al. (2014). Dynamics and associations of microbial community types across the human body. *Nature, 509,* 357–360.

21. Yurkovetskiy, L., et al. (2013). Gender bias in autoimmunity is influenced by microbiota. *Immunity, 39,* 400–412.

22. Koren, O., et al. (2012). Host remodeling of the gut microbiome and metabolic changes during pregnancy. *Cell, 150,* 470–480.

23. Sharon, G., et al. (2010). Commensal bacteria play a role in mating preference of *Drosophila melanogaster. Proceedings of the National Academy of Sciences, 107,* 20051–20056.

24. National Institutes of Health. (2015). Human Biome Project. Retrieved from http://nihroadmap.nih.gov/hmp

Chapter 8

1. Zerjal, T., et al. (2003). The genetic legacy of the Mongols. *American Journal of Human Genetics, 72,* 717–721.

2. Fisher, H. (2006). Romantic love: A mammalian brain system for mate choice. *Philosophical Transactions of the Royal Society B, 361,* 2173–2186.

3. Pease, A., et al. (1998). *Why men don't listen and women can't read maps.* Buderim, Queensland, Australia: Pease Training International.

4. Wlodarski, R., et al. (2014). What's in a kiss? The effect of romantic kissing on mate desirability. *Evolutionary Psychology, 12,* 178–199.

5. Hughes, S. M., et al. (2007). Sex differences in romantic kissing among college students: An evolutionary perspective. *Evolutionary Psychology, 5,* 612–631.

6. Ortigue, S., et al. (2010). Neuroimaging of love: fMRI meta-analysis evidence toward new perspectives in sexual medicine. *Journal of Sexual Medicine, 7,* 3541–3552.

Chapter 9

1. Zegers-Hochschild, F., et al. (2009). The International Committee for Monitoring Assisted Reproductive Technology (ICMART) and the World Health Organization (WHO) revised glossary on ART terminology. *Human Reproduction, 24,* 2683–2687.
2. Lynch, C. D., et al. (2014). Preconception stress increases the risk of infertility: Results from a couple-based prospective cohort study—The LIFE Study. *Human Reproduction, 29,* 1067–1075.
3. Wichman, C. L., et al. (2010). Comparison of multiple psychological distress measures between men and women preparing for in vitro fertilization. *Fertility & Sterility, 95,* 717–721.
4. Shindel, A. W., et al. (2008). Sexual function and quality of life in the male partner of infertile couples: Prevalence and correlates of dysfunction. *Journal of Urology, 179,* 1056–1059.
5. Nelson, C. J., et al. (2008). Prevalence and predictors of sexual problems, relationship stress, and depression in female partners of infertile couples. *Journal of Sexual Medicine, 5,* 1907–1914.
6. Peterson, B. D., et al. (2008). The impact of partner coping in couples experiencing infertility. *Human Reproduction, 23,* 1128–1137.
7. Peterson, B. D., et al. (2006). Gender differences in how men and women who are referred for IVF cope with infertility stress. *Human Reproduction, 21,* 2443–2449.
8. Jordan, C., et al. (1999). Gender differences in coping with infertility: A meta-analysis. *Journal of Behavioral Medicine, 22,* 341–358.
9. Glezerman, M. (1981). Two hundred and seventy cases of artificial donor insemination: Management and results. *Fertility & Sterility, 35,* 180–187.

Chapter 10

1. Visentin, M., et al. (2005). Prevalence and treatment of pain in adults admitted to Italian hospitals. *European Journal of Pain, 9*, 61–67.

2. Binkley, C. J., et al. (2009). Genetic variations associated with red hair color and fear of dental pain, anxiety regarding dental care and avoidance of dental care. *Journal of the American Dental Association, 140*, 896–905.

3. Lucey, P., et al. (2011). Automatically detecting pain in video through facial action units. *IEEE Transactions on Systems, Man, & Cybernetics, 41*, 664–674.

4. Wagner, T. D., et al. (2013). fMRI-based neurologic signature of physical pain. *New England Journal of Medicine, 368*, 1388–1397.

5. Craft, R. M., et al. (2004). Sex differences in pain and analgesia: The role of gonadal hormones. *European Journal of Pain, 8*, 397–411.

6. Riley, J. L. I., et al. (1999). A meta-analytic review of pain perception across the menstrual cycle. *Pain, 81*, 225–235.

7. Tolver, M. A., et al. (2013). Female gender is a risk factor for pain, discomfort, and fatigue after laparoscopic groin hernia repair. *Hernia, 17*, 321–327.

8. Cheung, C. W., et al. (2013). A large study assessing gender differences in postoperative patient controlled analgesia in Chinese population. *International Journal of Anesthesiology Research, 1*, 25–35.

9. Hoffmann, D. E., et al. (2001). The girl who cried pain: A bias against women in the treatment of pain. *Journal of Law, Medicine & Ethics, 29*, 13–27.

10. Alabas, O. A., et al. (2012). Gender role affects experimental pain responses: A systematic review with meta-analysis. *European Journal of Pain, 16*, 1211–1223.

11. Chen, E. H., et al. (2008). Gender disparity in analgesic treatment of emergency department patients with acute abdominal pain. *Academic Emergency Medicine, 15,* 414–418.

12. Interagency Pain Research Coordinating Committee. (2015). National pain strategy: A comprehensive population health-level strategy for pain. Retrieved from http://iprcc.nih.gov/docs /DraftHHSNationalPainStrategy.pdf

13. Wijnhoven, H. A., et al. (2006). Prevalence of musculoskeletal disorders is systematically higher in women than in men. *Clinical Journal of Pain, 22,* 717–724.

14. Ballantyne, J. C., & Sullivan, M. D. (2015). Intensity of chronic pain—The wrong metric? *New England Journal of Medicine, 373,* 2098–2099.

15. Hashmi, J. A., et al. (2013). Shape shifting pain: Chronification of back pain shifts brain representation from nociceptive to emotional circuits. *Brain, 136,* 2751–2768.

16. Ramirez-Maestre, C. (2014). The role of sex/gender in the experience of pain: Resilience, fear, and acceptance as central variables in the adjustment of men and women with chronic pain. *Journal of Pain, 15,* 608–618.

17. El-Shormilisy, N., et al. (2015). Associations among gender, coping patterns and functioning for individuals with chronic pain: A systematic review. *Pain Research & Management, 20,* 48–55.

18. Fillingim, R. B., et al. (2004). Sex differences in opioid analgesia: Clinical and experimental findings. *European Journal of Pain, 8,* 413–425.

Chapter 11

1. Ye, X., et al. (2012). Ambient temperature and morbidity: A review of epidemiological evidence. *Environmental Health Perspectives, 120,* 19–28.

2. Barnett, A. G., et al. (2005). Cold periods and coronary events: An analysis of populations worldwide. *Journal of Epidemiology & Community Health, 59,* 551–557.

3. Tipton, M. J. (2016). Environmental extremes: Origins, consequences and amelioration in humans. *Experimental Physiology, 101,* 1–14.

4. Zhong, C. B., et al. (2008). Cold and lonely: Does social exclusion literally feel cold? *Psychological Science, 19,* 838–842.

5. Byrne, N. M., et al. (2005). Metabolic equivalent: One size does not fit all. *Journal of Applied Physiology, 99,* 1112–1119.

6. Karjalainen, S. (2012). Thermal comfort and gender: A literature review. *Indoor Air, 22,* 96–109.

Chapter 12

1. Eine Krankheit namens Mann. (2003, September 15). *Der Spiegel, 38,* 150–159.

2. Statistisches Bundesamt. (2012, December 16). Statistik der natürlichen Bevölkerungsbewegung. http:www.gbebund.de

3. Luy, M. (2004). Causes of male excess mortality: Insights from cloistered populations. *Population & Development Review, 29,* 647–676.

4. Dinges, M. (2010). Männlichkeit und Gesundheit: Aktuelle Debatte und historische Perspektiven. In D. Bardehle & M. Stiehler (Eds.), *Erster Deutscher Männergesundheitsbericht: Ein Pilotbericht* (pp. 2–26). Munich, Germany: Zuckschwerdt Verlag.

5. Melamed N., et al. (2010). Fetal gender and pregnancy outcome. *Journal of Maternal-Fetal & Neonatal Medicine, 23,* 338–344.

6. Melamed, N., et al. (2009). The effect of fetal sex on pregnancy outcome in twin pregnancies. *Obstetrics & Gynecology, 114,* 1085–1092.

7. Neubauer, G., et al. (2010). Jungengesundheit in Deutschland. In D. Bardehle & M. Stiehler (Eds.), *Erster Deutscher Män-*

nergesundheitsbericht: Ein Pilotbericht (30–57). Munich, Germany: Zuckschwerdt Verlag.

8. Fabes, R. A., et al. (1994). The regulation of children's emotion regulation to their vicarious emotional responses and comforting behaviors. *Child Development, 65,* 1678–1693.

9. Krugman, S. (1995). Male development and the transformation of shame. In R. F. Levant & W. S. Pollack (Eds.), *A New Psychology of Men* (91–126). New York, NY: Basic Books.

10. Offner, P. J., et al. (1999). Male gender is a risk factor for major infections after surgery. *Archives of Surgery, 134,* 935–940.

11. Morales, A. (2004). Andropause (or symptomatic late-onset hypogonadism): Facts, fiction and controversies. *Aging Male, 7,* 297–303.

12. Bhasin, S., et al. (2011). Testosterone therapy in men with androgen deficiency syndromes: An Endocrine Society clinical practice guideline. *Journal of Clinical Endocrinology & Metabolism, 95,* 2536–2559.

13. Miao, H., et al. Incidence and outcome of male breast cancer: An international population-based study. *Journal of Clinical Oncology, 29,* 4381–4386.

14. Papaioannou, A., et al. (2008). The osteoporosis care gap in men with fragility fractures: The Canadian Multicentre Osteoporosis Study. *Osteoporosis International, 19,* 581–587.

15. Visram, H., et al. (2010). Endocrine therapy for male breast cancer: Rates of toxicity and adherence. *Current Oncology, 17,* 17–21.

16. National Comprehensive Cancer Network. *NCCN Clinical Practice Guidelines in Oncology Breast Cancer.* Retrieved from www .nccn.org/professionals/physician_gls/pdf/breast.pdf

17. Kessler, R. C., et al. (2003). The epidemiology of major depressive disorder: Results from the National Comorbidity Survey Replication. *Journal of the American Medical Association, 289,* 3095–3105.

18. Lyons, Z., & Janca, A. (2009). Diagnosis of male depression: Does general practitioner gender play a part? *Australian Family Physician, 38*, 743–746.

19. Addis, M. E. (2008). Gender and depression in men. *Clinical Psychology: Science & Practice, 15*, 153–168.

20. Martin, L. A., et al. (2013). The experience of symptoms of depression in men vs. women: Analysis of the National Comorbidity Survey Replication. *JAMA Psychiatry, 70*, 1100–1106.

21. Latalova, K., et al. (2014). Perspectives on perceived stigma and self-stigma in adult male patients with depression. *Neuropsychiatric Disease & Treatment, 10*, 1399–1405.

22. Lyons, Z., & Janca, A. (2009). Diagnosis of male depression: Does general practitioner gender play a part? *Australian Family Physician, 38*, 743–746.

23. Paulson, J. F., et al. (2010). Prenatal and postpartum depression in fathers and its association with maternal depression: A meta-analysis. *Journal of the American Medical Association, 303*, 1961–1969.

24. Ramchandani, P. G., et al. (2008). The effects of pre- and postnatal depression in fathers: A natural experiment comparing the effects of exposure to depression on offspring. *Journal of Child Psychology & Psychiatry, 49*, 1069–1078.

25. National Institutes of Health. (2000). Osteoporosis prevention, diagnosis, and therapy: NIH Consensus Development Conference statement. Retrieved from https://consensus.nih.gov/2000/2000osteoporosis111html.htm

26. Cummings, S. R. (2002). Epidemiology and outcomes of osteoporotic fractures. *Lancet, 359.* doi: http://dx.doi.org/10.1016/S0140-6736(02)08657-9

27. Johnell, O., et al. (2006). An estimate of the worldwide prevalence of disability associated with osteoporotic fractures. *Osteoporosis International, 17*, 1726–1733.

28. Pemmaraju, N., et al. (2012). Retrospective review of male breast cancer patients: Analysis of tamoxifen-related side-effects. *Annals of Oncology, 23*, 1471–1474.

29. Macdonald, H. M., et al. (2011). Related patterns of trabecular and cortical bone loss differ between sexes and skeletal sites: A population-based HR-pQCT study. *Journal of Bone & Mineral Research, 26*, 50–62.

30. WHO Scientific Group on the Assessment of Osteoporosis at the Primary Health Care Level. (2007). *Summary meeting report, Brussels, Belgium, 5–7 May 2007.* Geneva, Switzerland: World Health Organization.

31. Jiang, H. X., et al. (2005). Development and initial validation of a risk score for predicting in-hospital and 1-year mortality in patients with hip fractures. *Journal of Bone & Mineral Research, 20*, 494–500.

32. Calonge, N., et al. (2011). Screening for osteoporosis: U.S. preventive services task force recommendation statement. *Annals of Internal Medicine, 154*, 356–364.

33. Zvetov, G. (2014). Personal communication.

34. White, A. (2011). The State of Men's Health in Europe. Retrieved from http://ec.europa.eu/health/population_groups/docs/men _health_report_en.pdf

Chapter 13

1. Wichmann, M. W., et al. (1997). Male sex steroids are responsible for depressing macrophage immune function after trauma-hemorrhage. *American Journal of Physiology, 273*, 1335–1340.

2. Sykes, B. (2003). *Adam's Curse: A Future without Men.* London: Bantam.

3. Jones, S. (2002). *Y: The Descent of Men.* London: Little, Brown.

Chapter 14

1. Jameda. (2015). *Studie: Zwischen Wunsch und Wirklichkeit—Digitale Gesundheit in Deutschland*. Retrieved from http://www.jameda.de/presse/patientenstudien/_uploads/anhaenge/ergebnisprsentation_studie_digitale-gesundheit-6207.pdf

2. Megan, A., et al. (2014). Effect of a health system's medical error disclosure program on gastroenterology-related claims rates and costs. *American Journal of Gastroenterology, 109*, 460–464.

3. Eccles, R. (2002). The powerful placebo in cough studies? *Pulmonary Pharmacology & Therapeutics, 15*, 303–308.

4. Institute of Medicine. (2001). *Crossing the quality chasm: A new health system for the 21st century*. Washington, DC: National Academies Press.

5. International Allegiance of Patients' Organizations. (N.d.). Patient-centred healthcare. Retrieved from https://www.iapo.org.uk/patient-centred-healthcare

6. Epstein, M., et al. (2010). Why the nation needs a policy push on patient-centered health care. *Health Affairs, 29*, 1489–1495.

7. DeNoon, D. (2004, September 21). WebMD survey: The lies we tell our doctors. Retrieved from http://www.medicinenet.com/script/main/art.asp?articlekey=46985

8. Iezzoni, L. I. (2012). Survey shows that at least some physicians are not always open or honest with patients. *Health Affairs, 31*, 383–391.

9. Marvel, M. K., et al. (1999). Soliciting the patient's agenda: Have we improved? *Journal of the American Medical Association, 281*, 283–287.

10. Tannen, D. (2002). *You Just Don't Uunderstand: Women and Men in Conversation*. London, England: Virago.

11. Montagne, B., et al. (2005). Sex differences in the perception of affective facial expressions: Do men really lack emotional sensitivity? *Cognitive Processing, 6*, 136–141.

12. Kray, L. J., et al. (2014). Not competent enough to know the difference? Gender stereotypes about women's ease of being misled predict negotiator deception. *Organizational Behavior & Human Decision Processes, 15,* 61–72.

13. Nelson, A. (2004). *You Don't Say: Navigating Nonverbal Communication between the Sexes.* New York, NY: Prentice Hall.

14. Bertakis, K. D. (2009). The Influence of gender on the doctor-patient interaction. *Patient Education & Counseling, 76,* 356–360.

15. Roter, D. L., et al. (2002). Physician Gender effects in medical communication. A meta-analytic review. *Journal of the American Medical Association, 288,* 756–764.

16. Mast, M. S., et al. (2008). Physician gender affects how physician nonverbal behavior is related to patient satisfaction. *Medical Care, 46,* 1212–1218.

17. Derose, R. D., et al. (2001). Does physician gender affect satisfaction of men and women visiting the emergency department? *Journal of General Internal Medicine, 16,* 218–226.

18. Johnson, A. M., et al. (2005). Do women prefer care from female or male obstetrician-gynecologists? *Journal of the American Osteopathic Association, 105,* 369–379.

19. Fang, M. C., et al. (2004). Are patients more likely to see physicians of the same sex? Recent national trends in primary care medicine. *American Journal of Medicine, 117,* 575–581.

20. Bertakis, K. D., et al. (2009). Patient-centered communication in primary care: Physician and patient gender and gender concordance. *Journal of Women's Health, 18,* 539–545.

21. Schmittdiel, J. A., et al. (2009). The association of patient-physician gender concordance with cardiovascular disease risk factor control and treatment in diabetes. *Journal of Women's Health, 18,* 2065–2070.

22. Pickett-Blakely, O., et al. (2011). Patient–physician gender concordance and weight-related counseling of obese patients. *American Journal of Preventive Medicine, 40,* 616–619.

23. Lyons, Z., & Janca, A. (2009). Diagnosis of male depression: Does general practitioner gender play a part? *Australian Family Physician, 38,* 743–746.

24. Chan, K. S., et al. (2006). Does patient–provider gender concordance affect mental health care received by primary care patients with major depression? *Women's Health Issues, 16,* 122–132.

25. Bertakis K. D., et al. (2011). Patient-centered care: The influence of patient and resident physician gender and gender concordance in primary care. *Journal of Women's Health, 21,* 326–323.

26. Bertakis K. D., et al. (2011). Patient-centered care is associated with decreased health care utilization. *Journal of the American Board of Family Medicine, 24,* 229–239.

27. Association of American Medical Colleges. (2012). U.S. medical school applicants and students 1982–1983 to 2011–2012. Retrieved from https://www.aamc.org/download/153708/data/

Chapter 15

1. Women's Health Initiative (N.d.). WHI background and overview. Retrieved from http://www.nhlbi.nih.gov/whi/background.htm

2. Geller, S. E., et al. (2011). Inclusion, analysis, and reporting of sex and race/ethnicity in clinical trials: Have we made progress? *Journal of Women's Health, 20,* 315–320.

3. Zucker, I., et al. (2010). Males still dominate animal studies. *Nature, 46,* 690.

4. Hannah, M. E., et al. (2000). Planned caesarean section versus planned vaginal birth for breech presentation at term: A randomised multicentre trial. *Lancet, 21,* 1375–1383.

5. Glezerman, M. (2006). Five years to the term breech trial:

The rise and fall of a RCT. *American Journal of Obstetrics & Gynecology, 194,* 20–25.

6. Legato, M. J. (Ed.). (2010). *Principles of Gender-Specific Medicine*. London, England: Elsevier.

7. Rieder, A., & Lohff, B. (Eds.). (2008). *Gender Medizin. Geschlechtsspezifische Aspekte für die klinische Praxis*. Vienna, Austria: Springer Verlag.

8. Schenck-Gustafsson, K., et al. (Eds.). (2012). *Handbook of Clinical Gender Medicine*. Basel, Switzerland: Karger.

9. Oertelt-Prigione, S., & Regitz-Zagrosek V. (Eds.). (2012). *Sex and Gender Aspects of Clinical Medicine*. London, England: Springer Verlag.

INDEX